P9-CBL-424

KEY ENVIRONMENTS

General Editor: J. E. Treherne

MADAGASCAR

The International Union for Conservation of Nature and Natural Resources (IUCN), founded in 1948, is the leading independent international organization concerned with conservation. It is a network of governments, non-governmental organizations, scientists and other specialists dedicated to the conservation and sustainable use of living resources.

The unique role of IUCN is based on its 502 member organizations in 114 countries. The membership includes 57 States, 121 government agencies and virtually all major national and international non-governmental conservation organizations.

Some 2000 experts support the work of IUCN's six Commissions: ecology; education; environmental planning; environmental policy, law and administration; national parks and protected areas; and the survival of species.

The IUCN Secretariat conducts or facilitates IUCN's major functions: monitoring the status of ecosystems and species around the world; developing plans (such as the World Conservation Strategy) for dealing with conservation problems, supporting action arising from these plans by governments or other appropriate organizations, and finding ways and means to implement them. The Secretariat co-ordinates the development, selection and management of the World Wildlife Fund's international conservation projects. IUCN provides the Secretariat for the Ramsar Convention (Convention on Wetlands of International Importance especially as Waterfowl Habitat). It services the CITES convention on trade in endangered species and the World Heritage Site programme of UNESCO.

IUCN, through its network of specialists, is collaborating in the Key Environments Series by providing information, advice on the selection of critical environments, and experts to discuss the relevant issues.

KEY ENVIRONMENTS

MADAGASCAR

Editors

ALISON JOLLY

Rockefeller University, New York, USA

PHILIPPE OBERLÉ

Abidjan, Ivory Coast

ROLAND ALBIGNAC

Université de Besancon, France

Foreword by

HRH THE DUKE OF EDINBURGH

Published in collaboration with the

INTERNATIONAL UNION FOR CONSERVATION OF NATURE AND NATURAL RESOURCES

by

PERGAMON PRESS

OXFORD · NEW YORK · TORONTO · SYDNEY · PARIS · FRANKFURT

U.K.	Pergamon Press Ltd., Headington Hill Hall, Oxford OX3 0BW, England
U.S.A.	Pergamon Press Inc., Maxwell House, Fairview Park, Elmsford, New York 10523, U.S.A.
CANADA	Pergamon Press Canada Ltd., Suite 104, 150 Consumers Road, Willowdale, Ontario M2J 1P9, Canada
AUSTRALIA	Pergamon Press (Aust.) Pty. Ltd., P.O. Box 544, Potts Point, N.S.W. 2011, Australia
FRANCE	Pergamon Press SARL, 24 rue des Ecoles, 75240 Paris, Cedex 05, France
FEDERAL REPUBLIC OF GERMANY	Pergamon Press GmbH, Hammerweg 6, D-6242 Kronberg-Taunus, Federal Republic of Germany

First edition 1984

Library of Congress Cataloging in Publication Data

Main entry under title:
Madagascar.
(Key environments)
1. Zoology—Madagascar. 2. Botany—Madagascar.
I. Jolly, Alison. II. Oberlé, Philippe. III. Albignac, Roland. IV. Series
QL337.M2M33 1984 574.969'1 83-17394

British Library Cataloguing in Publication Data

Madagascar—(Key environments)
1. Natural history—Madagascar
I. Jolly, Alison II. Oberlé, Philippe
III. Albignac, Roland IV. Series
508'0969'1 QH195.M2

ISBN 0-08-028002-1

Printed in Great Britain by A. Wheaton & Co. Ltd., Exeter

BUCKINGHAM PALACE.

The general problems of conservation are understood by most people who take an intelligent interest in the state of the natural environment. But if adequate measures are to be taken, there is an urgent need for the problems to be spelled out in accurate detail.

This series of volumes on "Key Environments" concentrates attention on those areas of the world of nature that are under the most severe threat of disturbance and destruction. The authors expose the stark reality of the situation without rhetoric or prejudice.

The value of this project is that it provides specialists, as well as those who have an interest in the conservation of nature as a whole, with the essential facts without which it is quite impossible to develop any practical and effective conservation action.

Philip

1984

Foreword

Madagascar was probably one of the last regions discovered by man. Ever since the original fragmentation of Gondwanaland separated it from Africa, the Island remained an inviolate sanctuary for millions of years. There its unique fauna and flora could flourish, safe from all attack.

The first immigrants from Indonesia and Africa may have arrived at about the beginning of the Christian era, or perhaps somewhat earlier. These first Malagasy found a country covered with forests, and inhabited by strange animals: hippopotamus, Aepyornis, giant lemurs, carnivores and insectivores.

Flacourt, who visited Madagascar in the seventeenth century, left the first written description of the plants and animals of the Great Island. He described the astonishing character of Malagasy wildlife. However, human pressure was strong even then, and the largest and most archaic species, defenceless before man, were already becoming extinct.

Fortunately great treasure-lodes of Malagasy flora and fauna still survive. In the eighteenth century Commerson, marvelling, wrote the lovely phrases which introduce this book. In the nineteenth century Alfred Grandidier explored the Island, and undertook a monumental series of publications.

Many other scientists have continued the great work of discovering and describing Malagasy flora and fauna either as individuals or as representatives of research institutions. The work goes on today, as witnessed in the volumes of the Faune de Madagascar, the Flore de Madagascar, and many other erudite publications.

However, we still lack a good popular book where layman or expert can find a description in one volume of the essential characteristics of Malagasy nature. The present work therefore fills a major gap. It will allow us to better know, better love, better respect, and better protect the wildlife of Madagascar. The authors are all eminent scientists which guarantees the accuracy of this book, whose publication we salute.

The book appears at the right time, and deserves success. The Government of the Democratic Republic of Madagascar has recently authorized the creation of a Madagascar branch of the World Wildlife Fund. This decision presages major action to safeguard the Malagasy natural heritage during the years to come.

We thank Alison Jolly, Philippe Oberlé and Roland Albignac, editors of the book, as well as the authors. May "MADAGASCAR" contribute to world-wide appreciation of the importance and beauty of the plants and animals of Madagascar.

Vaohita

Barthelemi VAOHITA
Directeur de la Representation
du WWF a Madagascar.

General Preface

The increasing rates of exploitation and pollution are producing unprecedented environmental changes in all parts of the world. In many cases it is not possible to predict the ultimate consequences of such changes, while in some, environmental destruction has already resulted in ecological disasters.

A major obstacle, which hinders the formulation of rational strategies of conservation and management, is the difficulty in obtaining reliable information. At the present time the results of scientific research in many threatened environments are scattered in various specialist journals, in the reports of expeditions and scientific commissions and in a variety of conference proceedings. It is, thus, frequently difficult even for professional biologists to locate important information. There is consequently an urgent need for scientifically accurate, concise and well-illustrated accounts of major environments which are now, or soon will be, under threat. It is this need which these volumes attempt to meet.

The series is produced in collaboration with the International Union for Conservation of Nature and Natural Resources (IUCN). It aims to identify environments of international ecological importance, to summarize the present knowledge of the flora and fauna, to relate this to recent environmental changes and to suggest, where possible, effective management and conservation strategies for the future. The selected environments will be re-examined in subsequent editions to indicate the extent and characteristics of significant changes.

The volume editors and authors are all acknowledged experts who have contributed significantly to the knowledge of their particular environments.

The volumes are aimed at a wide readership, including: academic biologists, environmentalists, conservationists, professional ecologists, some geographers, as well as graduate students and informed lay people.

John Treherne

Contents

"May I announce to you that Madagascar is the naturalists' promised land? Nature seems to have retreated there into a private sanctuary, where she could work on different models from any she has used elsewhere. There, you meet bizarre and marvellous forms at every step....What an admirable country, this Madagascar"

(J-P. Commerson, 1771).

The aye-aye (Sonnerat, 1782)

Introduction

A. JOLLY, PH. OBERLÉ and R. ALBIGNAC

Ninety per cent of Malagasy forest species are unique to the Great Island. The aye-aye with its bat-ears, beaver-teeth, and skeleton finger is one of the world's rarest mammals, a relative of monkeys, apes, and ourselves. The piebald indri, large as a three-year old child, which sings in chorus from hill to hill, and the mouselemur, smallest of all primates, live with the aye-aye in Madagascar. Five scientific families of mammals, six of plants, and four of birds exist only here, including the thorn-studded spikes of Didiereaceae on the southern spiny desert, and the 12 species of Vangidae, which flutter and squawk through the dark rain forest of the east.

No list of species, however, conveys the scientific importance of Madagascar. This is a natural laboratory of evolution. Its origins are still disputed: perhaps it broke loose from Africa 100 million or perhaps even 180 million years ago. It was accessible to rafted colonists on logs or floating mats of vegetation until some 40 million years ago. Thereafter, its founding stocks have radiated in near isolation, to form that parallel world dreamed of by science fiction writers — and scientists. The rules of biology which still hold true in Madagascar are universal rules, for here the "ecological theatre and evolutionary play" has a different cast of characters.

And yet, today, 80% of Madagascar is covered by man-made prairie. This book began from a depressing paradox. During the years when the three editors lived in Madagascar, we realized all too clearly that for many tourists, and even Malagasy, the country seemed an ecological wasteland. We have too often heard visitors declare that the austere and silent landscapes of the plateau hold no attraction for them. Certainly the prairies, and the tracts of foreign pine and eucalyptus, which are fruits of patient efforts at reforestation, contain almost no animal life and few native plants. And far too often Malagasy have asked, with caution, "Is it true what people tell us, that lemurs exist only here?" or more bluntly, "Why don't you stay home and study your own lemurs in Europe or America?"

At the same time, we were part of the privileged group of naturalists who could travel to the rain forests, the dry deciduous woodlands, or the spiny desert, who have climbed the mountain peaks or burrowed in the caverns. With Malagasy and foreign colleagues we explored what Commerson called the "naturalists' promised land". Yet, because there are so few colleagues, we mainly talk to each other. There are library shelves full of scientific studies. They begin with Flacourt's *Histoire de la Grand Ile de Madagascar*, published in 1661. They continue with Alfred Grandidier's monumental 50 years of publication from 1866 to 1917, and include Perrier de la Bathie's 1921 account of *La Vegetation Malgache* which was both the first and one of the most lucid explanations of man's effect on the Malagasy landscape. Today, new volumes add to the current 136 of the Flore de Madagascar and the 60 of the Faune de Madagascar, for those with access to them. There is even an English technical overview (Battistini, R. and Richard-Vindard, G., 1972, *Biogeography and Ecology of Madagascar*, Junk, 1972).

In spite of the wealth of scientific knowledge, there have been only four accounts of Malagasy natural history in this century written for non-specialists. David Attenborough tells the funny and fascinating story of his travels there, while giving an excellent introduction to the wildlife (*Zoo Quest to Madagascar*, 1961). Raymond DeCary wrote a detailed catalogue, now long out of print (*La Faune Malgache*, 1950). Alison Jolly's *A World Like Our Own* (1980) described the plants and animals she watched on expeditions with Malagasy scientists to their own study sites. She tried to show how human needs interact with the needs of the forest, and the dilemma of conservation in a country where nature is rich but people poor. Finally, Philippe Oberlé compiled *Madagascar, Un Sanctuaire de la nature* (1981) including 103 colour plates of the Great Island. The chapters by Drs. Paulian, Guillaumet, Griveaud, Blanc, Benson, and Andriamampianina in this volume are translated from *Madagascar, Un Sanctuaire de la Nature* (Obtainable from Maison du Livre Specialise/Lechevalier, 7 rue Geoffrey Saint Hilaire, Paris 5e).

We are proud, then, to edit this handbook on Malagasy nature for English readers. We thank Dr. John Treherne and the Key Environments Series for the opportunity to do so. Above all, we hope that you, the reader, may now join the naturalists of the "promised land", and help to conserve the remnants of a fauna and flora unique in all the earth.

Alison Jolly
Philippe Oberlé
Roland Albignac

Etiene de Flacourt (1607 – 1660). His *Histoire de la Grande Ile de Madagascar* (1661) is the first great description of Malagasy natural history.

Rainilaiarivony, Prime Minister of the Merina Kingdom 1864 – 95. His "Code of 305 Articles" 1881, forbade felling virgin forest for charcoal or agriculture.

Alfred Grandidier, 1836 – 1921, author with his son Guillaume of the 34 volume *Histoire physique, naturelle, et politique de Madagascar*.

The Editors

Authors and Editors Addresses

Editors:

Dr. Alison Jolly
The Rockefeller University
1230 York Avenue
New York, NY 10021 USA

M. Philippe Oberlé
15 Rond Point Victor Hugo
92100 Boulogne
France

Professor Roland Albignac
Laboratoire de Biologie et d'Ecologie Animales
Faculte des Sciences
Universite de Besançon
Route de Gray
25030 Besançon
France

Authors:

M. Joseph Andriamampianina
Departement des Eaux et Forets
Universite de Madagascar
BP 175
Antananarivo
Madagascar

Mr. Constantine W. Benson (deceased)
communications to:
Mrs. C. W. Benson
Department of Zoology
University of Cambridge
Downing Street,
Cambridge
UK

Professor Charles P. Blanc
Laboratoire de Zoogeographie
Universite Montpelier III
BP 5043
34032 Montpelier Cedex
France

Dr. Leo H.M. Blommers
Herenstraat 102
3911 J H Rhenen
The Netherlands

Dr. Rose Blommers-Schlosser
Herenstraat 102
3911 J. H. Rhenen
The Netherlands

Dr. John F. Eisenberg
National Zoological Park
Smithsonian Institution
Washington, D.C. 20008
USA

Dr. Edwin Gould
National Zoological Park
Smithsonian Institution
Washington, D.C. 20008
USA

Dr. Paul Griveaud (deceased)
communications to:
Madame J. Griveaud
104 Av. du Littoral St. Marguerite
44380 Pornichet
France

Dr. Jean-Louis Guillaumet
INPA — Ecologia
C.P. 478
69000 Manaus
Brazil

Professor Richard Jolly
UNICEF
866 United Nations Plaza
New York, NY 10017
USA

Recteur Reynaud Paulian
La Rouviere
Port Sainte Foy
33220 Sainte Foy la Grande
France

Dr. Jean-Jacques Petter
Laboratoire d'Ecologie Generale
Museum National d'Histoire Naturelle
91800 Brunoy
France

Madame Rachel Rabesandratana
Universite de Tulear
BP 141
Tulear
Madagascar

Professor Guy Ramanantsoa
Banco School of Forestry
Abidjan
Ivory Coast

M. Barthelemi Vaohita
World Wildlife Fund — Madagascar
BP 4373
Antananarivo
Madagascar

CHAPTER 1

Madagascar: A Micro-Continent between Africa and Asia

RENAUD PAULIAN

The geographer E.-F. Gautier, at the dawn of the twentieth century, called Madagascar an island with the colour, the shape, and the consistency of a brick. A century and a half earlier, however, Philibert de Commerson described Madagascar as the naturalist's promised land. Scientists studying in Madagascar have fulfilled Commerson's prophecy by a wealth of discoveries. And yet, the contrast between these opposite summaries is enough to show that the amazing interest of Madagascar's fauna and flora may not leap to the eye of even highly cultivated visitors.

Promised land, certainly, but also a land which demands effort and perseverance. The effort itself only makes Madagascar even more fascinating: homeland of The Travellers' Palm, paradise of lemurs, microcontinent where immigrants from Indonesia have brought their language and their culture, where settlers from Africa and the Persian Gulf, India and even far-off Europe have succeeded in joining to create a Nation. The same effort imposed itself on all the European naturalists who first saw Madagascar from a flight over Nosy Bé southward to Tananarive, and who asked themselves in some anguish what they could possibly hope to accomplish in this naked eroded land. Many years later, at the moment of leaving, they have come to love every one of its aspects, unexpected or familiar. They have learned to appreciate the strange charm of its landscapes, and had the privilege of studying a flora and fauna unique in all the earth. The effort is, perhaps, even greater for naturalists born in Madagascar, working with very limited finance to combine the insights of two cultures — their own peoples' traditional knowledge of wild plants and animals with modern world-wide science. It is they who must confront the traditional view that nature is there for man's use, and set against it the twentieth century realization that all wilderness is fragile, perhaps too fragile to survive.

Madagascar lies near Africa's south-east coast, but if we can believe the geologists it broke off in the late Cretaceous from a more northerly position between Tanzania on the west and the Dekkan on the east. It then slid south and eastward toward its present position. This, at least, is the opinion of students of paleaomagnetism. A recent study shows, however, that the Mozambique Channel is composed of two identical basins which lie on either side of a longitudinal median ridge of sandstone. The sediments of these basins have been laid down with no discontinuities since the Cretaceous. This means that Madagascar forms part of the African continental plaque, and has occupied its present position since exceedingly ancient times.

Madagascar is escorted by a flotilla of isles and islets. On the map, all these Indian Ocean islands seem to form a homogeneous unit, and people commonly refer to a "Malagasy Region". In fact, one should distinguish five different island groups:

— a detatched fragment of ancient African basement rock, in the North: the Seychelles;

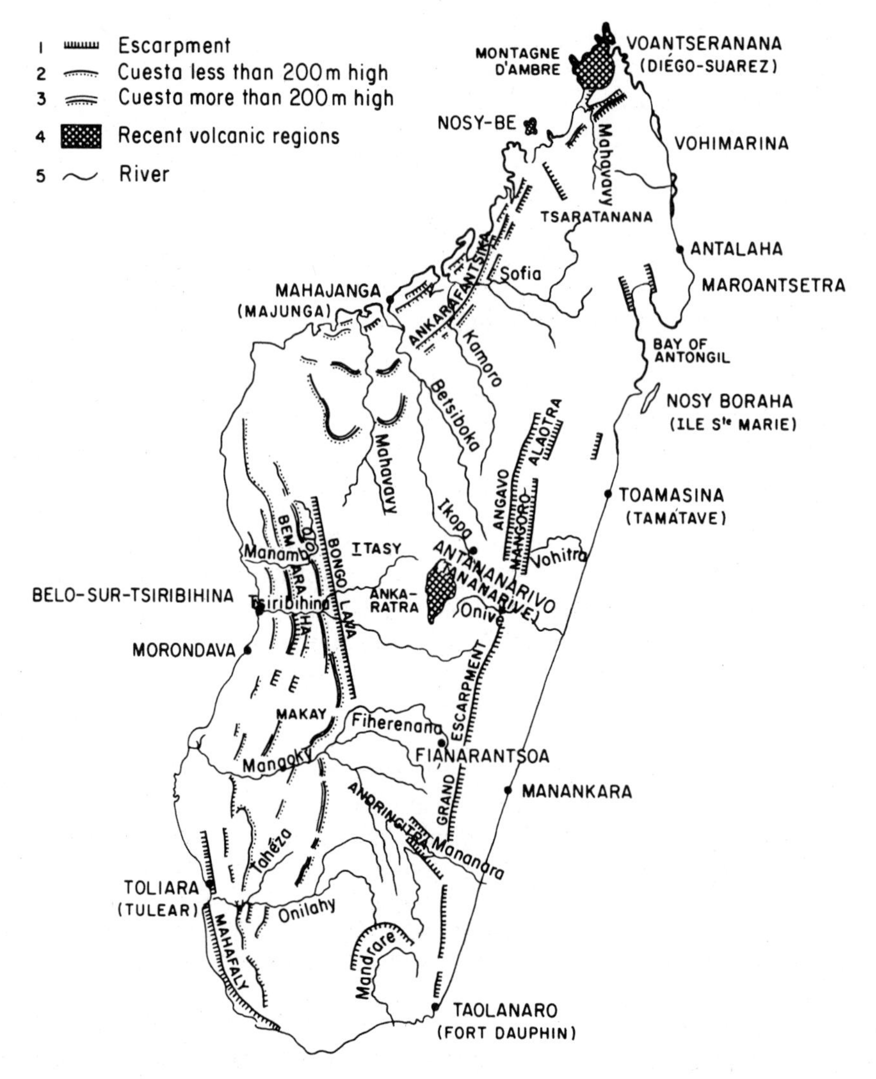

Fig. 1.1 Map of Madagascar (A. Jolly, after R. Battistini)

— two chains of volcanic islands on either side of Madagascar, on two fracture lines: the Mascarenes in a NE – SW line and the Comoros in an SE – NW line. In each of these groups the active volcanoes have shifted with time, westward in the Comoros and eastward in the Mascarenes;

— atolls or tiny islands built by corals in the shallows of the Mozambique Canal or the Indian Ocean: Europa, Juan de Nova, the Glorieuses, Tromelin, Aldabra, then, further off, the Amirantes and the Chagos Isles;

— finally, the Malagasy block itself, with the islands immediately offshore: Nosy Bé, Nosy Mitsio, Nosy Vorona, Nosy Ve, Ile Sainte Marie, Nosy Mangabe, and of course the islands in Diego Suarez Bay, as well as Nosy Lava, the Radamas and the Barren Isles.

This grouping reflects both palaeogeographic origins and faunal and floral divisions. Madagascar shares its plants and animals with the immediate offshore islands. The Comoros and Mascarenes, the distant atolls and the even more distant Seychelles have very different biotas, even if human contacts and relative nearness have led to some exchange of fauna and flora. Madagascar has certainly contributed colonizing species to Mauritius and Reunion, but only in recent historic times.

More detailed study of Madagascar itself reveals that it has all the characteristics of a continent, and must be treated as such. We are fully justified in concentrating on Madagascar and its offshore islands, without attempting to drag in the Seychelles and Comoros for a "regional description".

With a surface area of 587,000 km², and a North – South length of 1600 km, Madagascar is the fourth largest island in the world, after Greenland, New Guinea, and Borneo.

The Malagasy basement rock rests on a continental shelf which is larger than the island itself. On the east, contour lines are close together, and the sea reaches the great depth of 4000 m near the coast. On the south, shallower seas extend beyond the Red Island for nearly 1000 km. To the west the continental shelf is cluttered with coral reefs. It is fairly wide in places, but it drops off to deep sea, without joining the shelves round Europa in the centre of the Mozambique channel or the Comoro Islands further north. The Mozambique channel itself is nowhere deeper than 3000 m, and it is full of scattered shallows which rise singly or in clusters, often in the form of sharp peaks, or sea-mounts.

Madagascar has no high mountains, compared with neighbouring Africa: no summit rises above 3000 m. Some of these peaks do have typical mountain climate, but there is hardly any of Africa's montane ecological fascies. In spite of its generally low altitude the island's landscape is full of steep hills and valleys, created by the combined play of orogenetic phases with numerous fault lines that have sculptured the basement rocks and their overlying sediments, volcanic eruptions spaced out over time from the Cretaceous to Recent periods, and intense erosion resulting from climate and topography. It is interesting to note that the major fault lines follow the same direction as in Africa, with one series NNE – SSW, another NNW – SSE.

THE ANCIENT CRYSTALLINE BASEMENT

It is easiest to understand Madagascar's complex topography by geological classification. In doing so, we should recall our debt to Henri Besarie, founder and for many years Director of the Service Geologique de Madagascar. We owe to Besarie our basic geological knowledge of the Malagasy Region.

There are two main divisions: the ancient basement, and the halo of sedimentary rocks which border it.

The crystalline basement extends over 400,000 km², two thirds of the island's surface. It results from a major orogenic phase 2420 million years ago, which was doubtless followed by other large-scale orogeneses. The basement has been reworked in the course of time, with remarkable periods of mineralization — the oldest 2140 million years ago, formed the gold of the Vohibory, while the most recent, 484 million years ago, left deposits of thorionite. Between these two episodes the cipoline marbles of the central west were formed about 1890 – 1125 million years ago.

The great orogenetic phase of 2420 M. B.P. transformed three still earlier groups of rocks. The Androyan system is most ancient (the Behara granite, which cuts across the Androy was formed 3020 M. B.P., which puts it amongst the earth's oldest rocks). Then the Graphite system flows over the Androy granites toward the west, and this in turn still further west, is covered by the Vohibory system.

Metamorphism affected the ancient basement rocks with great intensity, and extended throughout the whole area: there are many facies with granulite and amphibolite. The basement, although cut by a double fault system, is so thick and so uniform that it perplexes geologists. The basement ends on the east by a straight-line fault of amazing regularity which forms the coast from Maroansetra to Fort Dauphin, and plunges on downwards to great sea depths of 4000 m.

Travelling inland, westward from the east coast, you come to an abrupt escarpment which rises in places 1900 m high. Then, beyond the escarpment there is a series of elongated rift valleys running north to south from the latitudes of Maroansetra to Fort Dauphin. The floor of these valleys lies only about 800 m above sea level; most of them have rivers at the bottom. The largest, the Mangoro, begins from

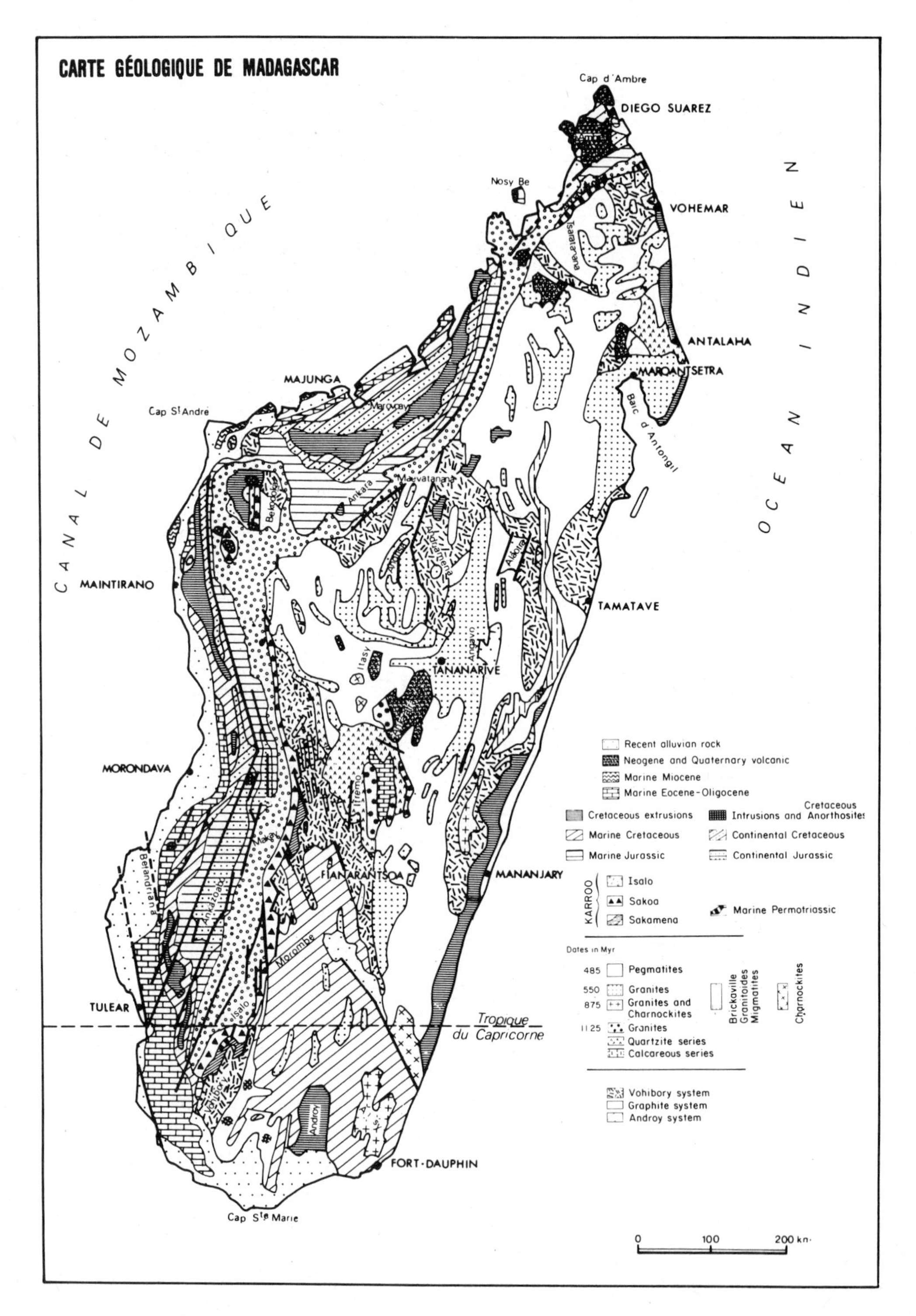

Fig. 1.2 Geological map of Madagascar (Atlas of Madagascar)

Fig. 1.3 Falls of the Mandrianampotsy on the Eastern Escarpment, between Fianarantsoa and Manakara (Ph. Oberlé)

the southern end of Lake Alaotra and flows southward to just east of Tsinjoarivo, then cuts through the escarpment cliff to descend to Mahanoro on the coast.

Still further west rise the more intricate chains of a second escarpment parallel to the first, that closes in the valleys. The inner escarpment reaches its highest point at 2650 m, the Pic Boby of the Andringitra massif. It is somewhat lower toward the south, reaching 1975 m in the massif of Andohahelo, and continues at around 2000 m to the north (for instance, Angavokely).

The massif of the Andringitra, in the centre of the second escarpment, is in its highest zones a chaos of granite, a rock desert with irregular growth of heath-like bushes. It has an extraordinary erosion pattern, apparently unknown elsewhere except in parts of Brazil. The granite domes and cliffs are covered with deep vertical channels called cunettes, visible from a great distance. No one seems to know how they are formed. To the northward beyond the Masoala peninsula, the ancient basement extends at low altitude, with hardly noticeable hills. Between the two regions a barrier of high, more recent mountains extends from the Marojejy (2135 m) on the east to the Tsaratanana (2876 m) on the west, cutting the island in two. Only narrow coastal strips offer a way round either side of this mountain chain.

West of the second or inner escarpment lie the High Plateaux. They extend from the mountain barrier to the north, southward down to the neighbourhood of Ambalavao. The High Plateaux contain three shallow basins: Ankaizina to the north, Imerina in the centre, and Betsileo to the south. Southward of Ambalavao the altitude drops from about 1200 to about 600 m. This is the Horombe plateau, which

Fig. 1.4 A bridge on the Eastern escarpment (Ida Pfeiffer, 1861)

forms the south–central region. It has a few highlands, of moderate altitude but great biological importance, such as the Analavelona and the Isalo, an extraordinary massif like the ruins of fortresses divided by deep canyons.

The basins form a depressed central zone covered with recent alluvions which came in part from erosion of the surrounding massifs and in part from volcanic eruptions between the Pliocene and Neogene. The sediments of the Lake Alaotra basin, in the valley between the two escarpments, have not given us any fossil animals, except a few aquatic insects in the upper layers of peat. The sediments in the Antsirabe–Sambaina basin, which have been particularly well studied, contain both Pliocene and Pleistocene deposits. In the Pliocene there were many fossil fish. In the lower Pleistocene we find crocodiles and herons, then a layer of ash contemporary with the last volcanic eruptions of the Ankaratra and then above them peat and clay with the subfossil fauna of giant lemurs, *Aepyornis* and *Hippopotamus*. The final layers of giant subfossils are contemporary with human settlement in the region.

Around the basins rise worn, rounded, naked hills, the *tanety*. Erosion scoops out huge gullies in the laterite layers on their flanks, the extraordinary *lavaka* so typical of the Malagasy landscape. On the sides of the tanety you may make out layers of pebbles in the soil. They are usually cobbles of rounded quartz, which doubtless reveal ancient erosion beds.

In many spots groups of rocks or huge granite monoliths, fragments of the ancient basement, loom over the tanety, giving the landscape a somewhat chaotic appearance. In fact, level table-like regions are relatively rare, in spite of the general name of "High Plateaux". There are a few flat areas, particularly in the west beyond the basins, which are locally called *tampoketsa*. Between the tanety, narrow ravines cut down through the laterite soil to bedrock. Torrents rush down them, over rapids and waterfalls.

Almost in the centre, the region is ruptured by the vast volcanic massif of the Ankaratra, and by the smaller but better preserved volcanic formations of Lake Itasy and its surrounding craters. Savannas cover the tanety and most of the hill-tops of the Ankaratra. Almost linear gallery forests survive at the bottom of ravines, and around the edges of swamps and peat-bogs. Local beliefs and taboos preserve

Fig. 1.5 The central highlands are bare savannahs dominated by isolated monoliths of basement rock. "Zazafotsy," or "the white-haired elder and the baby", near Ambalavao (A. Jolly)

a few groves on hill-tops, and these relicts are all that remain of the original continuous forest cover.

To the west of the High Plateaux, somewhat steeper hills are preceded by isolated low summits and then end in a well-marked escarpment running north to south. It drops away to the western lowlands, and it is called in general the "Bongo Lava" or "Long Mountain". It is an embankment rather than a cliff, and its slope is more regular and smoother than the cliffs of the eastern escarpment. All the same, there are a few ravines and hidden valleys where shreds of forest miraculously survive.

The red laterite soils of the High Plateaux have earned Madagascar its sobriquet of "The Red Island", as well as Gautiers' insult of "brick". Perhaps, though, there were other reasons for the name, perhaps even that the princely castes chose red clothes to express their rank. . . .

A MUSEUM OF MINERALS

Madagascar offers one of the world's most extraordinary collections of mineral deposits. Many types of rock were first discovered there, and bear the name of the Malagasy site where they were found. The diversity fascinates mineralogists. Alfred Lacroix consecrated three fat volumes to the description of the island's mineral riches. Unfortunately, in most cases we have hardly found more than samples. The deposits are so small, and so widely scattered, that they are only practical for small-scale artisans to exploit. Most of these rocks are silicates and niobates containing traces of rare earths.

Almost all the minerals are found in the crystalline basement, as eluvions, veins or inclusions. The exceptions found in other rocks include some phosphates in lower Cretaceous limestone, and more recent phosphates in Juan de Nova, coal, peat and lignite, asphalt and bitumen, a little raw copper in Cretaceous basalt, a bed of magnetite and a few veins of lead.

We can hardly list all of the minerals which occur in the crystalline rocks. Among them are: gold, silver, nickel, chrome, titanium, copper, lead, zinc, radioactive minerals (autunite mined at Vinanikarena, betafite at Ibity, euxenite, samiranite and ampaguebeite), cerium minerals (monazite, bastnaesite of the Imorona and tscheffkinite) haematite, ilmenite, industrial carborundum, zircon, rutile, thorianite, bismuth, piezoelectric quartz, mica, graphite. (The last two are actively mined.) These are all minerals for industrial use.

There are also many semi precious and precious gems, and Madagascar's wealth in jewels was for long a legend. Alphonse the Saintongeois already wrote of the riches of Malagasy "jewellery mines" in 1547. Flacourt, who was usually a more scrupulous observer, in 1658 cited topazes, aquamarines, emeralds, rubies and sapphires, mixing true and false.

The Merina kings forbade prospecting or use of gems. Ranavalona I expressly outlawed the search for gold, while Ranavelona II in 1881 issued an edict punishing mineral prospecting by 20 years in chains. This attitude may have derived from the religious interdictions of South-East Asia (Upper Burma), followed by the fear that any rich discovery might provoke a rush of foreigners into the country. It certainly delayed the study of Malagasy resources. The samples which E.-F. Gautier had studied in 1899 were the starting-point of serious work on Malagasy jewels, which nearly all come from pegmatites.

We should distinguish ornamental stones from semi-precious ones. The former include amethyst, rose quartz, lazulite, sun-stones and amazonites. Another group, found in the basalt, include calcedony, plasma, jasper, onyx and sardonyx. Celestite geodes from the Majunga region are superb collectors' pieces.

Among the true gems, a few small rubies and sapphires occur amongst the basaltic alluvia of the Ankarata. Many other gems are more frequent, including rubellite and many colours of other tourmaline, clear or pale blue topaz, light green spinel, chrysoberyl, orthose, danburite and scapolite, all golden yellow, green koonerupine, zircon, kunzite with a lovely rose-orange fluorescence under ultra-violet light which is only found elsewhere in some Californian stones, yellow garnet (spessartite) and deep red garnet and above all, beryl. The beryls are very diverse in density: light, with crystals reaching 30 kg and stones of 56 carats, or heavy with gems reaching 600 carats. They vary from blackish blue through green and yellow, and on to the peach-pink of morganite, a beryl unique to Madagascar. Beryls are the jewels of the Great Island.

These stones and jewels support artisans and traders who offer lovely minerals for export or to passing tourists. Collectors also buy polished slices of petrified wood and fossil ammonites, sawed in two with the cut faces polished. Exportation of fossil dinosaur bones, however, is forbidden, for the fossil beds were being pillaged and dinosaur vertebrae turned into tourist ashtrays!

As you might expect, hot springs and mineral waters are abundant throughout the ancient basement. They have been used by traditional healers, and a few sites, such as those at Antsirabe and Ranomafana, are also equipped for present-day "curists".

THE COASTAL PLAINS

Sedimentary rocks encircle the ancient basement rock. They lie in a narrow band along the east coast, and a wide halo around the north, west and southwest. The narrow coastal plain at the foot of the eastern escarpment includes a long line of Cretaceous volcanic rock, a few Maestrichian deposits, and a few continental sandstones which form low hills. Most of the plain, however, is recent alluvial layers left both by the rivers which flow from the eastern face of the inner escarpment and by erosion from the outer escarpment. There is so much of this eroded soil that it forms a band of barrier islands along the coast. Behind the barrier the rivers turn and flow north or south parallel to the coast, forming a swampy lagoon, until they reach some point of least resistance, or until a flood crest makes a temporary breach to the sea. This complex of water courses, lagoons and swamps has been largely joined together by man to form the "Canal des Pangalanes", a thoroughfare of travel and commerce.

Between the Pangalanes and the sea, the sandy barrier islands bear a forest of coconuts, filaos,

Fig. 1.6 The short rivers of the east coast flow across a narrow sedimentary plain. Soanirano, near Fort-Dauphin (A. Jolly)

Barringtonias and other colonizing trees. Even here there are endemic cycads, many Malagasy species of pandanus, and other indigenous species. Behind the Pangalanes, the low plain and hills were the domain of a great coastal forest which has almost disappeared today.

In the west of the island the sedimentary rocks are far more widespread and complex. They derive from several geologic stages: primary, most from the secondary and to a lesser degree from the tertiary eras.

A first group of formations occupy a low lying zone, a sort of trough which lies along the basement rocks at the foot of the Bongo Lava. In the north, and for a shorter stretch along the Vohibory in the south, this valley is occupied by Permotriassic marine deposits. In the centre of the trough lies a narrow band of deposits called the Sakamena and the Sakoa which run from the Sakoa river to the level of Maintirano. Tillites and black schists form the lower layers of the Sakoa, followed by layers of coal covered by sandstones.

The Permotriassic valley is bounded on the west by areas of high relief, composed of Isalo sandstone which forms ruiniform massifs in places, and dates from the middle Triassic to the Bathonian. West of these Primary rocks we find Jurassic and Cretaceous marine and continental deposits. At first they form a more or less continuous rocky barrier: the Makay, the Antsingy, the Ankarana. They continue as a plain, either wide and open or broken into scattered blocks (Namaroka). They are riddled with grottos and caverns through which underground rivers flow, the home of cave fish and blind prawns. Recent explorations have revealed a network of underground galleries more than a kilometre long in the Ankarana cliff.

The south and west show, in places, vast stretches of recent soils, which include petrified wood on a Pliocene terrace, laterites, probably of Neogene origin, and red sands. The wide coastal plain undergoes a brutal shift between the wet and dry seasons, in the calcareous regions. The soil is hard-baked, dry and dusty in the dry season, but during the rains it is transformed into thick, sticky mud, into which

Fig. 1.7 Wind-carved Jurassic sandstones of the Isalo Massif (A. Jolly)

Fig. 1.8 Gorges of the Manambolo, below the Bemaraha Plateau (Ph. Oberlé)

you may sink to the knees, which somewhat impedes travel at this period.

The amount of erosion in this region can be shown by observations made in the Betsiboka estuary. During the first fifty years of this century, the Betsiboka deposited 40 metres thickness of sediment, or nearly a metre per year.

Fig. 1.9 The calcareous rock of the Mahafaly plateau bears drought-resistant "spiny forest". The Onilahy River, unlike many rivers of the southwest, never dries completely (A. Jolly)

Fig. 1.10 Caverns of Anjohibe or Andranoboka, 82 km from Majunga. The network of galleries is 2 km long, 50 hectares in area (Ph. Oberlé)

THE FOSSILS, WITNESS OF AN EXTINCT FAUNA

Just as the basement rocks are remarkable for their minerals, so the sedimentary layers are remarkable for their fossils. Paradoxically, in spite of man's late arrival on the island, people coexisted with, knew and doubtless contributed to the disappearance of the giant subfossils. This subfossil fauna is one pole of attraction for palaeontologists in Madagascar. The other lies at the other end of the geologic time, in the fossils of the primary era.

Most of these subfossil remains date from 2200 to 2300 years ago, which must correspond to a critical period for these creatures, probably linked to climatic change.

Freshwater and ponds which once covered parts of the High Plateau dried up at that time, leaving only a few relicts such as Lakes Itasy and Alaotra. The immense lake basins of Antsirabe and Betafo were reduced to insignificant marshes. The giant vertebrates which had reached their maximum size could not adapt to the new conditions. They first became rare, then disappeared, like the dwarf hippopotamus whose remains were discovered near Antsirabe. Human actions, joined to the climatic change, contributed to some animals' extinction. We have found no trace of prehistoric man on Madagascar. The sum of present knowledge indicates man arrived on the Great Island in the first centuries of the Christian Era, in successive waves of colonization from the African coast and Indo-Malaya.

The first immigrants probably came from the coastal villages of East Africa, people who already combined the Indonesian, Dravidian, and pre-bantu blood which characterized the coast. Later, "islamic" towns were implanted in the northwest, northeast, and a few points on the east coast of the island.

These islamic towns for a period knew real splendour. They were trading stations and shipping harbours, with buildings of coral blocks, which had large cemeteries, and graves holding tripodal cooking-pots made of chloritoschist hollowed on a lathe. Their influence must have stretched fairly widely, for we find chloritoschist pots of the same style on the southeast coast and in Anjouan on the Comoros.

The chloritoschist statue of Vohitsara is undoubtedly a carving of the extinct dwarf hippopotamus. Louis Mollet thinks that it was carved where it was found, and that it cannot be an elephant as other authors claim.

A legend spread at this time through all the Arab seafaring world: a legend of a giant bird named the Roc. Even Marco Polo reported it in his travels, and said it lived on an island to the south called Madagascar. It was probably Aepyornis, the "elephant bird", exaggerated even beyond its own size by sailors' imagination.

The Ratites are an order of heavy running birds who cannot fly. Apart from extinct species, like Madagascar's *Aepyornis* and *Mullerarnis*, they include the African ostriches, the South American rheas, the cassowaries of New Guinea, Australian emus and the Kiwis of New Zealand.

Alfred Grandidier discovered the fossil remains of *Aepyornis* in 1868, at the bottom of a pond near Tulear. The bird could grow to 3 m in height and a weight of 450 kg. Flacourt in the seventeenth century stated such birds still lived in the "Ampatra woods", according to the inhabitants of the south.

A fairly large number of *Aepyornis* eggs, which hold seven or eight litres, have survived intact; some sand dunes are riddled with shell debris. It was a fairly defenceless prey so both the birds and their eggs were much hunted by the first inhabitants, just as in New Zealand where the Maoris hunted another ratite group, the Moas, to extinction.

Man also knew the last giant lemurs: *Megaladapis*, *Archaeoindris* and *Paleopropithecus*. They were heavy animals, perhaps more clingers or walkers than climbers, for they could only move slowly in the trees. The largest were the size of a gorilla.

Flacourt again following local accounts, describes a strange beast. "The tretretretre is an animal large as a two-year-old calf, with round head and human face, but both fore and hind feet like a monkey's. It has frizzy fur, a short tail, and ears like a man's. One has been seen by the pool of Lipomami, which ranges in that neighbourhood. It is a very solitary animal; the people of the country fear it greatly and flee from it as it flees from them."

This description could apply to a fantastic creature out of Malagasy legend, or could equally well be a giant lemur. The imposing skeleton of a squat and powerful *Megaladapis* is on exhibition at the Museum of the Academie Malagache. The head, which is highly compressed laterally measures 30 cm long. The powerful upper teeth include two long curved canines which are both pointed and shearing, strong pointed premolars, and enormous molars. In the lower jaw, behind the small, narrow and nearly horizontal "toothcomb" of incisors, a large premolar forms a formidable "scissor" against the upper canine. The animal was equipped to devour fruit, leaves, and even tough branches or roots. The large pelvis recalls that of an ungulate, but the long, robust fingers allowed the animal to climb trees. Even if men feared it, they also hunted it, as proved by traces of blows or gashes on several fossil bones.

Beginning in ancient times between the first and tenth centuries, and perhaps as late as the seventeenth century for a few of the hardiest or best protected species, a large part of the Malagasy fauna has disappeared. It was one of the world's unique and divergent faunas, but it has left us nothing but fossils and legends.

On the other hand, the giant terrestrial tortoises, larger even than the modern tortoises of Aldabra or the Galapagos, probably disappeared earlier towards the end of the Pleistocene. Finally as regards the giant crocodiles called "*Crocodilus robustus*" described from Pleistocene deposits, they were simply huge specimens of the modern species *C. niloticus*. Even in the last thirty years we have seen the average size of crocodiles greatly diminish due to intensive hunting of the largest specimens.

When we look beneath the abundant subfossil layers, fossils from the tertiary era are rare on the whole. There are only a few sites from the Miocene and the Oligocene. The Eocene is better represented and offers a great variety of sea-urchins, Milioles, Nummulites, and calcareous algae of the genus *Limnothalamnion*.

In the secondary era Malagasy paleontology again reveals remarkable creatures, of the sort the subfossils led us to expect. The sediments are largely marine, with an abundance of diverse ammonites, as well as reptiles. At the beginning of the secondary a rich flora accompanies the remains of fish, while in more recent strata there are dinosaurs of huge size.

The most remarkable Malagasy fossils, however, date from the earliest periods, the Carboniferous and the Permian. It was then that the first marine transgressions began to separate Madagascar from southern Africa. There are of course, analogies between the primary faunas of Madagascar and South Africa, for example, reptiles of the genus *Tangasaurus*, common to both. In general, however, the two reptile faunas differ profoundly. Madagascar had only *Sauropsidae*, the group which gave rise to both reptiles and birds. The South African fauna is basically *Theropsidae*, the ancestors of mammals. This offers an explanation of the unique nature of Madagascar's mammalian fauna. There is a total absence of ruminants, elephants, large carnivores, and monkeys. The mammals which populated Madagascar were accidental arrivals, rafted across the Mozambique Channel at rare intervals. Once arrived they have speciated explosively *in situ*. We should also note that South Africa underwent a period of intense volcanism in the Primary which Madagascar escaped, a difference of great importance in interpreting the past history of the region.

The Primary fauna of Madagascar includes Brachiopods, Ammonites, sea urchins, the reptiles, very handsome fossil fish in the schists between Beroroha and Malaimbandy. Above all, extraordinary nodules of fossil fish of the lower Triassic have accumulated in an immense cemetery which stretches from Ankitokazo to the Mahavavy. This rich and beautifully preserved fauna is related to that of Greenland and Angola, but has no equivalent in South Africa. It is a marine fauna, but laid down close to shore for plant imprints associate with the fish as well as Stegocephalus, a terrestrial *Protobranchus*, and rare terrestrial insects of exceptional interest.

The Carboniferous has left some 560 million tons of coal, which is full of cinders and of little industrial use. It resembles the coal of India and South Africa, called Gondwanian Coals. It contains important plant fossils including *Glossopteris* which survived until the Permotriassic. Unfortunately for the biogeographer there are no insect remains.

Below the coal strata, the sandstones contain few fossils. At least they have yielded one Theropsid, and there are calcareous beds with Brachipods.

A VARIED CLIMATE

Madagascar undergoes an enormous diversity of climates. This may seem surprising, for it all lies at fairly low altitude. The Piton des Neiges, in the tiny island of Reunion, rises higher than any Malagasy summit.

The diversity derives in part from Madagascar's geographic position between 12° and 25° south latitude, which ranges from subequatorial humid weather to subtropical dry. It derives far more though, from the north – south direction of the principal mountain ranges, which thus lie perpendicular to the prevailing winds. The Windward Zone is exposed to the humid breath of the Monsoon, and receives most of the rainfall brought from the Indian Ocean. It is sharply divided from the Leeward Zone, which only receives a little left over moisture from the east winds. It can scarcely get much rain from the west, for the west winds have blown over Africa and their passage over the Mozambique Canal is too short to recharge them with moisture.

The great length of the island and the rugged relief multiply climatic differences into a real mosaic of conditions. The simple latitude differences, to start with, mean that even two seaside sites will have appreciable temperature differences ranging from 27° annual mean at Diego Suarez to 23° at Fort Dauphin.

The annual temperature cycles at Diego Suarez have a typical equatorial pattern of two annual maxima in December and March. As you progress southward this changes gradually to a temperate pattern with a single annual maximum in January, at Fort Dauphin.

Seasonal rains complicate the picture on the west coast where we find two maxima and two minima as far south as Majunga. The great rains of December result in a secondary temperature drop.

The two hottest regions of the island lie along the north west coast, and at Bemaraha, which reach average maxima of 34°. As you gain altitude, the temperature falls rapidly: the yearly average at Tananarive is 17.3° as against 24° at Tamatave. On the high summits there are occasional low temperatures and frost, which have resulted in some tragedies during their exploration. Minima of – 15° and snow which remains on the ground occur sometimes on the Andringitra. Besides this, the daily temperature range is far greater on the mountains.

Rainfall varies even more than temperature. The highest annual rainfall of 3612 mm/year falls on the Ile Sainte Marie; the least falls at Anakao, a mere 310 mm/year. Erratic yearly changes increase this difference even further. On the whole, rainfall decreases from east to west, and from the latitude of Ile Sainte Marie it decreases both northward and southward. These differences occasionally jump from one extreme to another. In the extreme south you only travel over some 50 km from a rainfall of 533 mm/year at Behara to 1529 mm/year at Fort Dauphin!

For naturalists, climate cannot be analysed simply as a function of extreme values or of variations in temperature and rainfall. A climate chart combines these with the temporal distribution of heat and moisture. The biological effects of climate depend as well on the soils and their capacity to hold water or permit evaporation.

Following Morat we can distinguish five main categories of climate in the island: perhumid, humid, subhumid, semi-arid and subarid. Several of these have distinct variants according to latitude or altitude, so we end with eleven different climatic types.

Perhumid climate characterizes the east coast from Antalaha to Fort Dauphin, as well as the eastern

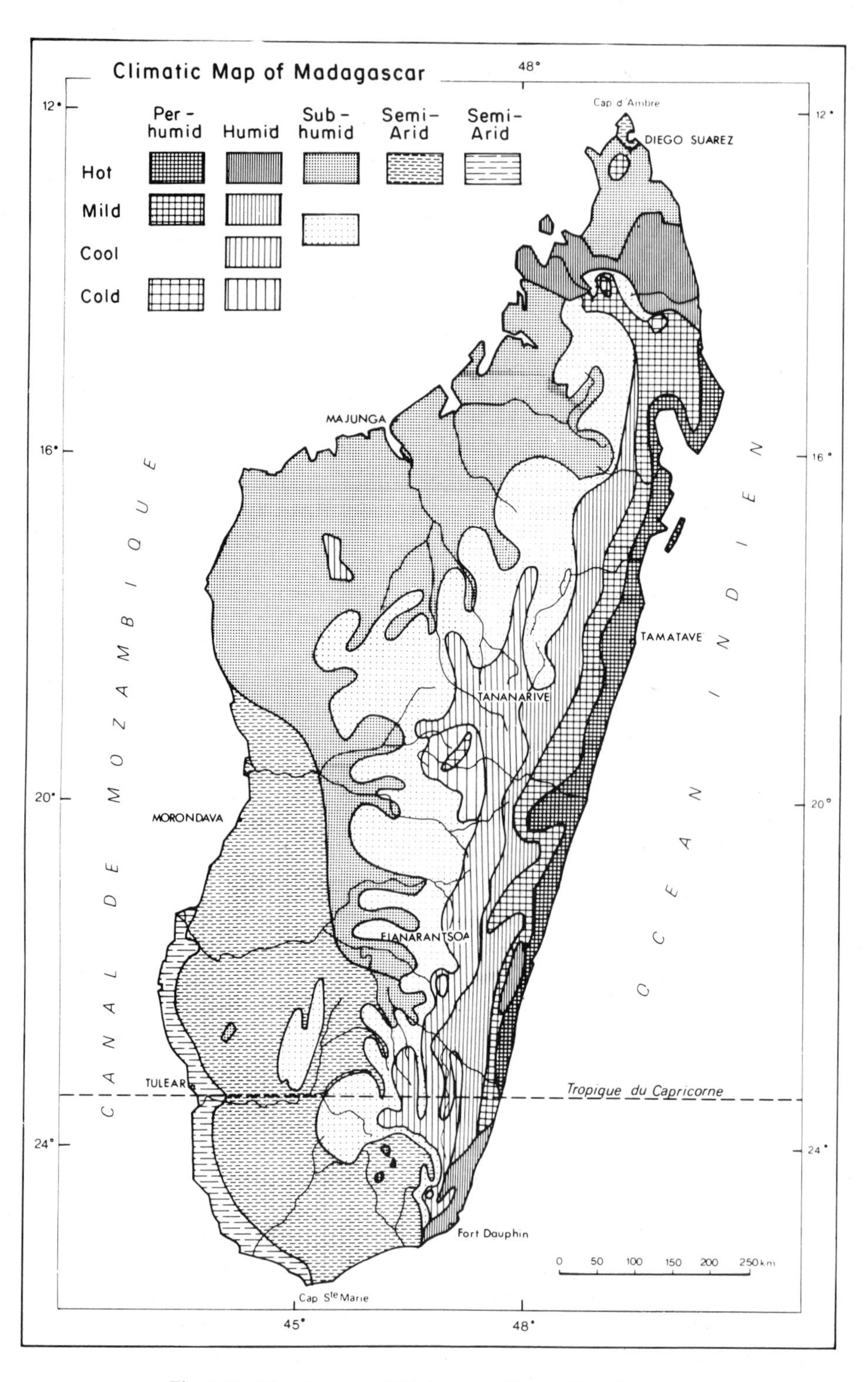

Fig. 1.11 Climatic map of Madagascar (Atlas of Madagascar)

escarpment, the summits of the Ankaratra and the Montagne d'Ambre and the mountain chain which runs from east to west between Maroansetra and the Sambirano. These regions have rainfall greater than 1850 mm/year, with a dry season less than a month long or sometimes nonexistent. At high altitudes with this climate there may be rainless seasons of three months or more, but clouds and fog maintain high humidity throughout the year. Annual temperature is relatively constant, averaging from 15.5° – 24° according to altitude.

Perhumid climate divides into three distinct types: hot, temperate, and cold. The first is found at low altitudes. The temperate type lies higher, starting at 400 – 800 m in the north but only 30 m above sea-level in the south. The cold type reigns above 1600 m on the Ankaratra and above 1800 m on the Marojejy or the Tsaratanana.

Humid climate covers the region from the Sambirano to Vohemar, the interior escarpment and the eastern part of the plateaux. Rainfall varies between 1000 to more than 2000 mm/year, with a well-marked dry season lasting from 3 – 6 months. The dry season is sharpest where the temperature is highest. Mean annual temperature varies with altitude and latitude between 14° – 26°, with larger yearly fluctuations than under the perhumid climate.

Humid climate includes four types: hot (on the coast), temperate, cool (at Tananarive) and cold (that of Antsirabe).

Subhumid climate has rainfall from 950 mm – 1600 mm/year, with a clear-cut dry season of 4 – 7 months. Mean annual temperature varies from 17 – 28°, with sharp annual and daily fluctuations according to altitude.

This climate dominates the north and west in its hot type, and all the western plateaux and highlands of the south in its temperate form.

Semi-arid climate has a rainfall between 500 – 900 mm/year, and a 7 month dry season, at the least, with strong annual variation. Mean annual temperature varies from 25 – 26°. This is the climate of the southwest, the south, and the Cap d'Ambre Peninsula. In this region has been recorded the absolute maximum temperature for the island: 44°.

Subarid climate, has an average rainfall of 350 mm/year, falling so irregularly one can scarcely speak of a wet season, but only of occasional rains. Mean annual temperature is about 26° with maxima of around 40°.

This highly specialized climate occurs in a narrow strip along the coast, roughly some 30 km wide, from Morombe to Tulear and round to Cap Sainte Marie. This region is noted for the weird Malagasy bush with its Euphorb trees and its Didiereaceae, growing on calcareous crust or on red sands. The Didieraeacae also grow in the semi-arid region of the south, but they are characteristic of the subarid climate.

At Cap Sainte Marie this woody vegetation is flattened by the wind into a horizontal carpet of interlaced trunks at ground level.

LAKES AND RIVERS

Madagascar possesses many crater lakes of which some, like Lake Tritriva, are famous for their beauty, and there are many ponds and little stagnant pools. However, Madagascar has only five large lakes. Two lie on the High Plateau: the immense stretches of Lake Alaotra where open water blends with marsh, and the great Lake Itasy. Three lie on the west coast: Lake Kinkony near Majunga, Lake Ihotry in the lower Mangoky region and Lake Tsimanampetsotsa on the edge of the Mahafaly Plateau. Lake Ihotry shelters many water birds. Its variable level makes it a favoured area for nesting and raising nestlings, and a sanctuary for a fauna highly threatened elsewhere. Lake Tsinamampetsotsa is salty and has an impoverished fauna except where fresh water springs flow on its east coast, attracting pink flamingos.

Fig. 1.12 The falls of the Lily River in the volcanic region of Itasy (Ph. Oberlé)

The Malagasy rivers fall into three groups. On the east coast narrow streams leap down the escarpment as torrents, rapids and waterfalls, then make their way to the sea across the coastal sediments. On the cliffs they run in rocky beds. Aponogetonaceae grow in their clear waters, particularly *Aponogeton fenestralis* or *Ovirandrana*, whose long strap-shaped leaves are latticed with regular rectangular openings. In the upper reaches of the streams live little endemic fish, some of which like *Bedotia* are excellent aquarium species. There are specialized insects: The rapids contain *Hydrostachrys*, aquatic plants which nourish the caterpillars of a huge moth. These streams have seasonal floods, which are sometimes devastating after cyclones, but they also have a large runoff all year round.

Their innumerable waterfalls and rapids are jewels of the Malagasy landscape, sometimes on a grand scale. The rapids of the Betsiboka near Maevatanana are stained red with silt, then are engulfed by narrow ramparts where flecks of spray bespatter the tops of the highest trees on the bank. The northern highway bridge straddles the site so a traveller can look straight down at the falls.

The waterfalls of the Mahavavy near Ambilobe, and the great cascade of the Montagne d'Ambre are also justly famous. The falls of the Mandrianampotsy are visible from the Fianarantsoa-Manakara railway.

Those of the Sakaleona, the highest in Madagascar with a drop of 200 m, demand a small trip from Nosy Varika.

Two great rivers of the northwest, the Mahavavy and the Sambirano resemble the eastern streams in their upper courses. They are born near the sea but at high altitude on the Tsaratanana. They reach the plain in a few steep kilometres of cascades. Even at low water they run strongly and their outlets are unstable. The other rivers of the west originate on the western face of the inner escarpment. They form a fan enlarging toward the west as they cross the high Plateau, drop over the Bongo Lava, and finally traverse the western plain. We can list them from north to south: the Manongarivo, the Maevarano, the Sofia, the Mahajamba, the Betsiboka, the Southern Mahavavy, the Ranobe, the Manambolo and the Tsiribihina which cuts through the Bemahara massif in narrow canyons, the Morondava, the Mangoky, and finally, in the southwest, the Fiherenana and the Onilahy. All these rivers take slow, majestic courses across the plains, with only a few mill-races and an occasional spectacular waterfall. They flow into the Mozambique Canal. There are violent flood crests in the rainy season, particularly after cyclones, but very low water in the dry season. Every year the shrinking floods leave vast areas of fertile silt, the *baiboha*, ready for dry season crops (tobacco, cotton, ground nuts, etc.).

Finally, three rivers flow southward into the Indian Ocean: the Linta, the Menarandrana and the Mandrare. They originate in the western buttresses of the Ivakoany and the eastern escarpment. They have occasional floods in the rainy season and a large underground flow throughout the year. However, in the dry season, they may become beds of dry sand where people have to dig pits to find water for themselves and their herds.

NATURAL LANDSCAPES IN CONSTANT RETREAT

Man first arrived in Madagascar less than two thousand years ago, as dated by Carbon^{-14} tests. The early, sparse settlements were limited to the coastal plains for several centuries. Nevertheless, man the latecomer has modelled the landscape, adding his stamp to the action of geology, relief, and climate. The low population density gives the human effects a special character.

The first effect was deforestation, doubtless accelerated by the slow drying which results, according to Perrier de la Bathie, from the progressive erosion of the north – south mountain range.

Recent studies have confirmed that when men first arrived the island was covered with closed forest. There was high lowland rainforest on the east coast, humid forest at the middle and high altitudes hung with epiphytes and with bare or grass-covered soil, and montane rainforest on high windy summits, with dwarfed, twisted trees, covered with moss and lichens and growing out of a dense carpet of moss. On the High Plateau and western buttresses of the escarpment there was humid forest interspersed with corridors of deciduous woodland, ericoid bush on a few summits, and dry deciduous woodland which began on the High Plateaux and continued over the western and northwestern plains. Further south grew the "bush", a strange flora of spiny trees and bushes, especially Didiereaceae and Euphorbs. On the western coast, thick mangroves grew in the shoreline mud.

The only openings in this forest cover were occasional lakes with their own vegetation, a few peat-bogs on rocks, here and there a peak or vast rock with their succulent rock-plants, and finally grassy clearings left from the forest fires lit by lightning. The clearings were temporary, open for a time, then regrown.

Large-scale climatic change from wet to dry, accompanied by rising and falling sea-level, have certainly affected the Malagasy forest cover. The advance and retreat of different plant communities explains the occasional unexpected relict patches, or a species found apparently far from its proper home. These movements have also contributed to Madagascar's extraordinary multiplication of species by leaving behind genetic isolates.

However, nothing leads us to think that climatic change could give a natural explanation for the vast open savannahs before man's arrival. Instead, fire did its work all too quickly around each human settlement. It was probably used at first to clear a little land for cultivation and to protect the village from surprise attack. Later it became a systematic tool to open new ground for shifting cultivation leaving exhausted soil after the crop for forest regrowth. Fire grew in importance when it was used to clear or "maintain" pasture for the ever-increasing herds of humped cattle. These cattle, commonly called zebus, were introduced to the island by the first inhabitants, along with another breed lacking humps, which has only recently disappeared. In tribal Malagasy social structure, the size of the herd indicates, or even is, the rank of its owner. During religious feasts and funerals hundreds of cattle are sacrificed, and the tomb is decorated with their horned skulls. In this system the number of cattle matters far more than their individual quality. Fire opens new pasture ground to the cattle, destroys forest covers used by brigands and cattle-thieves, and helps limit parasites, particularly ticks. Finally if the fire is cleverly timed, it is followed by early regrowth of grass during the dry season, the "green bite" which carries the herds through to the rains.

Ever since the tenth century every autumn from north to south and from the eastern escarpment to the Mozambique Canal, the days are filled with curtains of smoke on the wind, and the nights with the red glow of bush fires, across the Malagasy landscape.

The destruction was relatively slow at first, but accelerated over the centuries. We know that Mayeur in the mid-eighteenth century travelled from Tananarive to Ambositra still walking under forest. A century later Imerina lay bare and bald beneath the heavens, to the point where its people were nicknamed "Ambaniandro", "those who live beneath the sky". In the market of Tanananive grass replaced wood as cooking fuel.

In spite of desperate efforts by the department of Eaux et Forets the destruction is still accelerating today: bush fires supposed to "maintain" the pastures, *tavy* or slash and burn rice cultivation, and now the industrial exploitation of forest for its wood, and forest land for plantation crops. Every year the edges of forest blocks, gallery forests bordering rivers or huddled in ravines in the bare savannah, the groves round certain peaks or tombs which were protected by religious beliefs or respect for the ancestors, are laid low by the flames.

And yet, in spite of the increasingly dry climate, the forest regains lost ground wherever it is protected. We have proved this both on the western plains, in the Reserve of Namoroka, and in the high ground of the east, on the Soaindrano plateau of the Andringitra massif. Photos taken of the same spot over fifty years' interval clearly show the degree of reforestation.

The Malagasy have long been aware of forest regrowth, after its destruction, wherever it is protected, and have also known that second growth differs profoundly from primary forest. They use two different words: *ala* means primary forest, and *savoka* regrown forest.

The secondary woodland or savoka is a mixture of light-demanding, rapidly growing trees. Their light or yellow green leaves cut across the dark green of the virgin forest.

Very, very slowly, the savoka in turn is replaced by primary forest — if the fire does not come again. There are always seeds left in the soil, or brought from neighbouring plots. Unfortunately the original fauna may not return: it has been shown that the large terrestrial carabid beetles of the genus *Scarites* have not recolonized a particular forest zone one whole century after its destruction, though the vegetal cover has regrown to all appearances just as before.

This underlines the importance of the eleven Reserves Naturelles Integrales. These are wholly inviolate reserves, set aside in 1927, which makes Madagascar a pioneer among the countries of the tropics in wilderness legislation. Both in their size, their quality, and the scientific studies made within them, these reserves are crucial in forest preservation. There are, as well many smaller special reserves, and parcels of forest conserved through local beliefs, which also play a major role.

Human change in the vegetation began with deforestation and the replacement of forest by grasses.

Later, as fires repeatedly burned the plains, grasses themselves have disappeared. In parts of the south central region the ground has the appearance of a steppe of hard beaten earth, here and there sprouting meagre tufts of *Aristida*, which is virtually unusable by man or beast. Only termites eat *Aristida*, pimpling the surface with their small roughly conical termite hills, in which they accumulate short segments of the grass cut to uniform length.

Before reaching the final stage of barren pseudosteppe, the savannahs go through an intermediate phase with a few trees — isolated or in tiny clumps.

Along the watercourses, the woodland cover remains as gallery forest with *Pandanus* and *Ficus*, which survive far longer than the other trees. The gallery forest can shrink to a thin cordon of trees or even a single file of *Pandanus* or palms, standing aligned in profile against the sky.

Deforestation speeds erosion by wind and rain. The superficial layer of soil, no longer firmly held, simply disappears. With it goes all the forest fauna which lived from leaves, fruit or wood, which sought shade or humid soil: lemurs and birds, and also insects, molluscs, frogs. The Malagasy fauna being rich in highly localized species, even local forest destruction entails the final extinction of many species, and the impoverishment of the genetic pool which might have offered useful animals or plants to man.

Accelerated erosion leads to disastrous floods, and the silting up of lowlands along the coast. In the long term we can even foresee a general flattening of relief, and lowering of the great north – south chain, thus hastening the general desiccation of the island.

Fig. 1.13 Lavaka erosion (E. Gould)

Fig. 1.14 Over most of Madagascar, natural forest has been cleared, leaving sterile land. The plateau west of Tananarive lies bare to grass fires and erosion (A. Jolly)

MAN'S MARK ON THE GREAT ISLAND

Man has changed the landscape through general deforestation. He has also remodelled it on a smaller scale, by creating innumerable rice fields on the High Plateaux.

The immigrants, masters of Asian techniques of irrigated rice culture, have sculpted every hillside, every montane amphitheatre, every stream head and spring into the regular stair steps of paddy terraces. Their colours vary with the seasons from the brilliant yellow-green of the little square nursery fields; the *ketsa*, to the forthright green of the replanted seedlings, to the pale blonde of high stalks bearing the harvest as they await the sickle.

The highest and loveliest rice fields recall the paddy terraces of the Philippine mountains. They lie on the hillsides of the Betsileo country around Ambositra and Fianarantsoa. The terraces are sometimes too narrow even for a shovel, and can only be tilled with an *angady*, the narrow bladed tool which the peasant holds with both hands and plants in the ground with a loud ''ugh''.

Under the forest cover of the East coast, the rice terraces which hang on the ravine sides are irrigated by conduits made of sections of giant bamboo stems.

Houses dot the landscapes. On the east coast and the escarpment, and to some extent in the west and the Sambirano, houses are made of woven panels of palm fibres or bamboo splints. The floor stands on stilts, well above the ground, and the roof is thatched with leaves of palm or travellers' palm.

These light huts, made of *falala*, are airy and healthy. They are also built by Sakalava and Mahafaly peoples. The Vezo, a fishing tribe of the southwest, are more likely to throw a pirogue sail over a tripod of mast and spars, which resembles an American Indian teepee.

In the Antandroy country, traditional huts are extremely small, often too low for an adult to stand upright inside. They are built of straight planks of the light *Alluaudia* wood (fantsilotra). Like the larger houses of Ambohimanga they were decorated with rich carvings.

The houses are usually grouped in small villages aligned along a path or a canal.

In Imerina, the central province, the disappearance of the trees forced people to construct houses of beaten earth from very early times. Wood became the prerogative of princely dwellings and royal palaces. House timbers were carried on men's backs, or simply dragged by man-power over long distances from the eastern forest. The wheel only reached the Plateaux in the nineteenth century, though it was known three centuries earlier on the coast. A few houses in Tananarive, including the one which sheltered Jean Laborde as well as the palace-house of Andrianampoinimerina show us the craftsmanship and audacity of those Malagasy carpenters. Among the Zafimaniry, forest people of the Ambositra region, the tradition of wooden houses persists today.

In early times, mud houses were built by treading clay then piling it up in thick layers to form housewalls which dried in place. They could even build two storey houses in this fashion. The process was slowly improved: first by using huge blocks, then hand-made bricks, then moulded sun-dried bricks, and finally oven-firing. Clay-pits and kilns dot the Plateaux landscapes. A bright red slip made of clay, cowdung and banana-juice covers and protects the outer walls.

The most primitive houses held one small room. Two-storey houses succeeded them, often crowned by a loft-granary, and then provided with a verandahed balcony on the two long sides. These houses are common in the Betsileo country, usually with an outside staircase. In Imerina the staircase is inside, between the two main rooms of the ground floor.

Nowadays, sadly, the red outer coating is replaced by diverse colours which blend uneasily with the red landscape, and the ancient forms disappear.

Until recently the Plateaux villages were built on hill tops well above the rice fields. This was mainly for security reasons, and perhaps also to check malaria. They were surrounded by an outer dry moat and a circular or oval wall, the doorway blocked by an enormous disc of stone. The protection created by these fortifications was not negligible. One of the tales retold in the *Tantara*, the History of the Merina Kings, tells of a King whose village was beseiged by superior forces. He ordered the dry moat filled with rice straw, set fire to it and covered it over with a thin layer of earth. His unsuspecting enemies attacked, sank, and perished in the burning mass.

The Merina monarchy brought security. The walled villages burst apart into separate farms and large domains. Agricultural buildings were set on the hillsides; farmhouses, granaries, chicken houses, storehouses and a big kitchen garden all enclosed by a rectangular wall of mud blocks.

Later still, with the law and order of colonization, the development of commerce and the end of slavery, even the rectangular wall was no longer needed. Little houses are set like a string of beads at the foot of the hills beside the rice-fields, or along the road to market.

Human settlement has profoundly modified the larger basins of the High Plateaux. These are generally crossed by a river, and are flat enough for large-scale rice growing. They are crisscrossed by a dense network of canals, full of locks, dikes, and tiny bridges. As the rivers gradually deposit silt, they have built up their beds until they flow above the level of the plain. The protective dikes which contain them are large as the dikes of Holland.

The farmers of these regions have water in abundance. Like the farmers of the west who till seasonal swamps, the *matsabory*, they use cattle to trample and fertilize the paddy-field before sowing. Peasants often combine fish farming and rice: carp, introduced at the turn of the century by Dr. Legendre, goldfish introduced earlier, *Tilapia* imported by the Service de Eaux et Forets after the Second World War. They prosper in the fields and canals. People fish with lines, nets, and by the simple expedient of draining a field. These fish are the chief source of protein for ordinary Malagasy.

Finally, people have changed the High Plateaux over the past three-quarters of a century by an active policy of reafforestation. Groups of grey-green Eucalyptus and plantations of darker green pines cover and protect a part of the soil, preparing for the future. The first plantations were destined for firewood: Eucalyptus to fuel the Tamatave – Tananarive railway. Now pines are grown for building wood, paper pulp, and to protect the soil against erosion.

Toward lake Itasy, plantations of Aleurites look like industrial fields, while in the Mangoro valley plantation manioc also geometrizes the landscape. Other commercial agriculture squares out the plains of the Samibirano, the west, and Antandroy, respectively with plantations of sugar cane, cotton and tobacco, and sisal. Agriculture on the seasonal floodplains of the west, the baiboho, imposes a large-scale but less geometric pattern on the great river valleys. Other crops do not strike the eye so clearly; those which are planted under cover of shade-trees: cocoa, coffee, peppers, and vanilla.

Only the plantations of ylang-ylang of the northwest and Nosy Bé are totally bizarre. The ylang-ylang branches are systematically pruned or broken downwards until the short, tortured trees resemble gnarled pines on windy mountains near the timber line. Their flowers are distilled for perfume. The odour of ylang and coffee flowers greets the traveller in season, long before he can see the coast of Nosy Bé.

Finally, man has placed his mark on the landscape by building tombs. The Malagasy everywhere share similar beliefs in the power and importance of the ancestors. In spite of this unity of traditional religion, tombs in the different regions take very different forms.

On the East Coast there were common open tombs where men and women lay separated, as well as closed wooden coffins, often in the shape of pirogues, grouped together in the forest.

In the West, the Sakalava tombs of Menabe included a wooden enclosure surmounted by carved figures, whose origins clearly trace back to ancient Asian sources. These realistic figurines show people, couples and birds. Some of these are certainly pelicans, which proves there was a colony of these birds somewhere in the region.

In the south the Mahafaly built virtual fortresses as tombs: square platforms 10 – 15 m on a side, and 1 – 1.5 m high. They were surmounted by 4 – 16 *aloalo*, carved geometric posts topped by a figurine — usually a zebu, but in recent times bush-taxis or even aeroplanes. The skulls of zebu sacrificed at the funeral lie on the tomb. Southern tombs are isolated from each other, and often placed beside roads, with a fantsilotra planted at each corner. On the High Plateaux tombs also sprinkle the hills: stone tombs in Betsileo, masonry ones in Imerina.

If habitations, tombs, agriculture, and irrigation have contributed to the landscape in their various ways, it is not yet so for industry. Only a few spoil-heaps and trenches reveal the presence of mines. More and more hydroelectric dams relegate the lovely waterfalls to the past, like those of the Mandraka or the Namorona. Artificial lakes appear, like Tsiazompaniry or Mantasoa. Mantasoa, lying in a many-armed valley, even offers a basis for leisure sports.

Urban development appears in the vitality of many towns, centres of counties or provinces, markets, ports. Rural exodus and shanty-town suburbs are less of a problem in Madagascar than in many other tropical countries: small towns and villages still hold their attraction.

If you fly over Madagascar at low altitude, the Great Island appears like an archipelago, a constellation of populated areas separated by vast stretches of mountain, forest, and above all the bare prairie kingdom of wandering cattle herds. A thin network of tracks connects the islands of habitation. This fundamentally divided nature of the country, as well as its broken relief, imposes constraints on social, economical and political development.

Another dominant trait of Madagascar is the beauty of its light, the extraordinary variety of its atmospheres and the impression of serene melancholy which hovers over its landscape, above all on the High Plateaux.

From the dramatic scenery of the crater lake Tritriva, to the softness of the emerald ricefields, the

Malagasy landscape gives off infinite nuances. Malagasy are profoundly aware of their landscape — tradition and local proverbs endow lakes and trees with their own significance. We shall not forget the attitude of the porters, during one expedition in the mountains, when they profited by every halt to sit facing out over the vast spaces, bathed in a sublime light, and awaiting in peaceful meditation the time to depart again.

ORIGINS OF THE FAUNA

The unique character of the Malagasy fauna and flora and its profound differences from that of neighbouring Africa, led naturalists to begin asking very early what links existed between the island and other continents. The enigma of Madagascar did not just stimulate botanists and zoologists, but geologists and geophysicists as well.

The originality of the Malagasy fauna and flora derives from the high level of endemism, from the diversification of particular groups which are rare or unknown elsewhere, from the richness in archaic forms, and also from the astonishing gaps.

This is not the gradual impoverishment of an insular fauna. Instead it is the evolution in a closed community of disparate elements derived from ancient Gondwanaland and from accidental later arrivals.

Depending upon which groups you consider, there are clear African affinities (for most of the fauna), Indo-Malaysian affinities (for example among Molluscs and certain frogs and bats), or divided between the two poles (for birds). A southern element which reappears in South America, South Africa, Australia and New Zealand is feebly but clearly represented.

Naturalists have imagined a number of different primeval continents to explain this distribution. Lemuria would have been an Indo-Malagasy continent, Gondwana a vast southern continent, Sudamadie a South-African – Malagasy land. Plant and animal biogeography alone cannot prove any particular continental links. At most, probable diffusion routes and probable barriers are suggested. Too many factors have affected species distribution on the earth's surface for it to be otherwise.

Geology and geophysics allow surer paleogeographic reconstructions. Geology indicates that the first division between Madagascar and Africa occurred at the end of the middle Permian (the Sakoa formation) and at the base of the upper Permian. The relative isolation of Madagascar from South Africa at that period is proved by the lack of vulcanism in Madagascar while it was intense in Africa, and by the differences already noted in the fossil reptile faunas.

Later Madagascar became ever more isolated, until it was totally cut off in the Cretaceous. Perhaps tertiary changes in the Mozambique Canal were large enough for island chains to appear, offering stepping stones for immigration by some African species.

The basement rocks are geologically like those of South India, but Madagascar's separation from that subcontinent predates the division from Africa, in spite of some hypotheses to the contrary. A major volcanic episode occurred on both sides of the island during the Cretaceous, with large flows of basalt.

The ancient Malagasy basement is very like that of both South India and South Africa, and has undergone similar evolution up to the end of the Isalo stage. We can, therefore, speak of a Gondwanaland region without prejudging the role of South America and Australia.

Geophysicsts, from the evidence of paleomagnetism and the nature of the earth's crust, affirm that Madagascar was initially situated between the east coast of Africa at the level of Tanzania, and the west coast of peninsular India. It then slid south and east relative to Africa, into its present position. India, instead, moved north and much farther east. But recent geological evidence differs (see p. 2).

It is interesting to add, on this subject, that the volcanic "hot spots" of the Comoros, South Africa, and Madagascar have always moved westward, or else the land eastward, during a given period.

In conclusion, we can tentatively unite Madagascar to a group of ancient southern lands, and even a Gondwana, without being overprecise about the near geographic relations of these lands, and certainly without exaggerating their importance in the origins of fauna and flora. Most of the modern plant and animal groups had not evolved at the time of Madagascar's separation from India and Africa. Except for a few very old Gondwanaland forms, they could only reach the island after its isolation, by rare accidental transport.

Just like Australia, a continent evolved in complete isolation since before the evolution of placental mammals, so Madagascar has evolved as an island since before the first mammals appeared. Sea currents, floating vegetation, cyclones and winds, and transport by migrant birds have brought Madagascar the elements of its fauna and flora since that early period.

Now, in modern times, a group of anthropophile species are arriving with humans by sea and air. They are often hardy invaders which menace the more fragile endemics. Voluntary introduction of carnivorous fish, black bass and trout, are an added danger to the indigenous fauna.

When the original species reached this land, they found a nearly empty region, with low selection pressure. The diversity of topography and climate cut it into a multitude of regions, all distinct in their ecology. The combined play of this feeble selection pressure, genetic inheritance, and extreme diversity of milieux, has permitted exceptional speciation. This has happened in two very different fashions:

— on the one hand, great morphological and ethological diversity, as among the lemurs, rodents and insectivores. Descendents of a single ancestor are adapted to many different "ecological niches".
— on the other hand, a multitude of morphologically distinct forms which replace each other geographically in the same ecological niches. This explosive speciation by "pulverization", happens in many groups. Among the insects we could cite among the most spectacular and best known forms: Canthonines, Curculionids of the genera *Proictes* and *Neseremnus*, and Anchomenian Carabids. It even happens among underground fauna and is clear among the blind Osorians recently studied by Coiffait. Thus, we may characterize the Malagasy fauna as very isolated, with many gaps, retaining many archaic traits, and originating largely from Africa though with some rare southern and oriental elements. Isolation and low selection pressures at the start favoured an enormous speciation among those forms of life which could find an initial foothold. Thus, in a single fauna we find the insular traits of archaism and faunal gaps, combined with the continental trait of intense speciation.

This evolution continues before our eyes, making Madagascar a natural laboratory of extraordinary fascination.

CHAPTER 2

The Vegetation: An Extraordinary Diversity

JEAN-LOUIS GUILLAUMET

Which should we admire most in the Malagasy flora: its amazing endemism, or its bizarre and unexpected patterns of growth? It is certainly a unique flora, with 12,000 species of flowering plants, of which 85 per cent are confined to Madagascar. They are members of 180 plant families, of which 6 or 7 are endemic, and nearly 1600 genera, a quarter of which are endemic.

This numerical richness is not the only reason for the fascination of the Malagasy flora, but it poses many problems to botanists: the origins, the affinities (whether African, Asian, pantropical or still other) and the plants' diversification by region or locality. Active present day speciation gives us material for interpreting the grand sweep of life, and raises immediate questions of economic use. Thus, the extreme diversity of Malagasy wild coffees might open new ways to improve this universal beverage; the variety of Madagascar periwinkle species may reveal an untapped source of medicines. The island was so recently settled that it has no true indigenous cultivated foods, but many wild plants are gathered for food or remedies. The inventory of these natural resources, which has gradually grown over the years, should be enlarged and backed by analyses and experimentation for practical human benefit.

The second aspect, no less important because it leaps to the eye, is the attractiveness of Malagasy plants: their beauty of form and colour, their strangeness at times, their constant originality. Madagascar has given few plants to the World's gardens but they are among the jewels: the flamboyant (*Delonix regia*) from the dry forests of the West, the traveller's palm (*Ravenala madagascariensis*) from the deforested hillsides of the east, or again *Chrysalidocarpus lutescens*, an elegant palm of the eastern forest. We should not forget the humble Madagascar periwinkle (*Catharanthus roseus*), whose extracts check childhood leukaemia in the continents of the north.

Sharp ecological variations and the physical separation of regions have favoured the diversification, creation and explosive speciation of so many taxa. These factors are also responsible for the geographic mosaic of vegetal cover. We can pass in a few hundred metres from humid forest to dry bush, from monotonous prairie to shady, luxuriant forest groves. Man has not economized Madagascar's original riches; he has too often abused and squandered his natural heritage.

To gain a vivid impression of the Malagasy vegetable kingdom let us travel across the different regions of the island, observing the main traits of each landscape and its special peculiarities, and noting on the way a few plants remarkable for their scientific importance or their beauty. Most of these plants are unknown outside Madagascar; they have no common names in other languages. In Malagasy, a single species may be called differently in different regions, or the same name apply to more than one species. Besides, the Malagasy names are newer to most readers than the Latin ones!

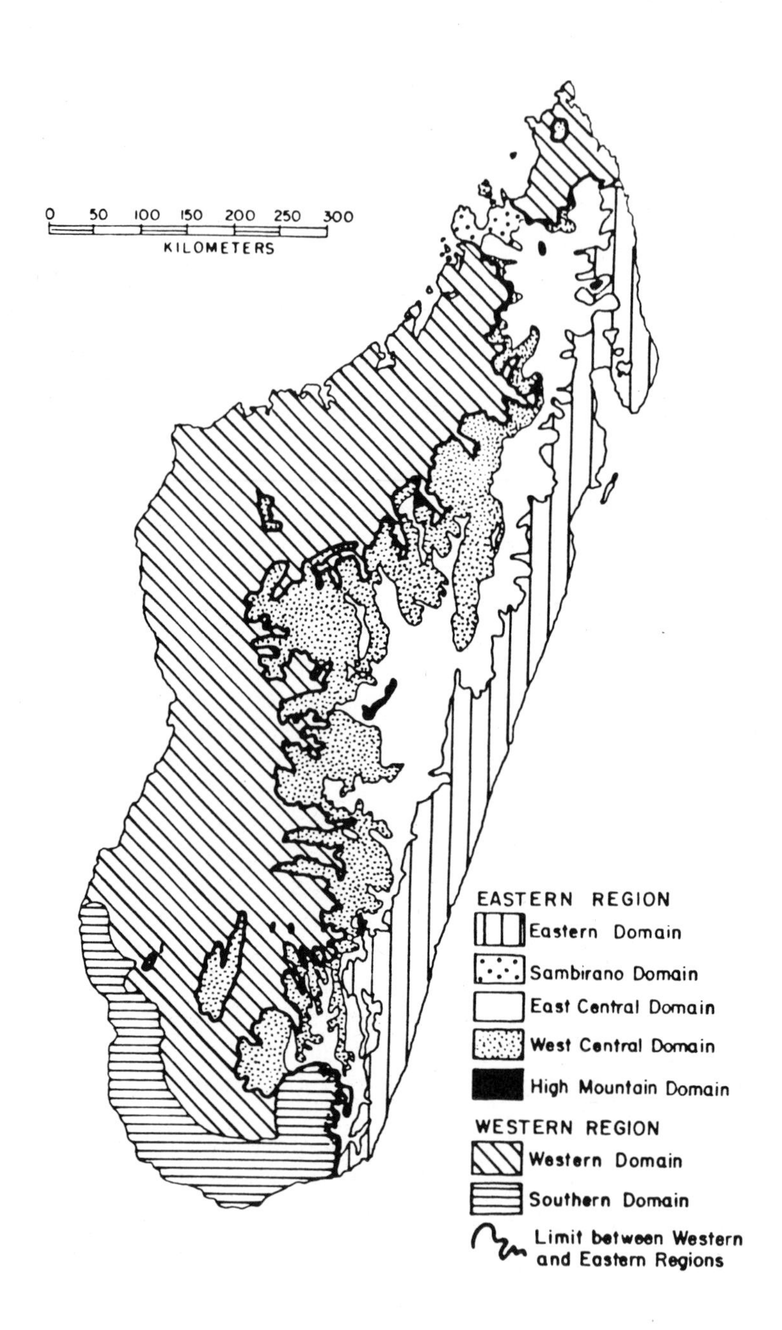

Fig. 2.1 Climax vegetation types, as established by H. Humbert (1955), after Tattersall (1981)

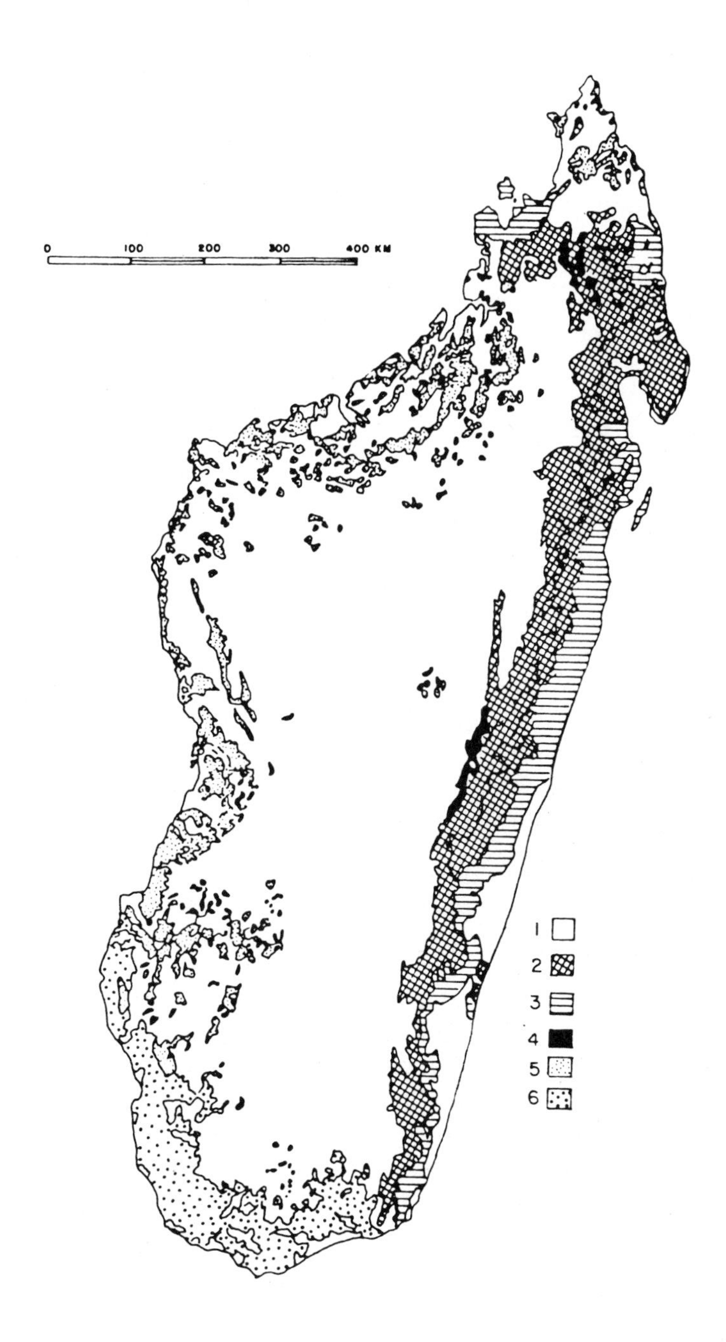

Fig. 2.2 Remaining forested areas, from Eaux-et-Forets survey 1949 – 58 after Humbert and Cours Darne 1965, and Tattersall 1981. 1. savannah and steppe; 2. dense rain forest; 3. savoka (secondary humid forest); 4. montane forest; 5. deciduous woodland; 6. xerophilous bush (spiny desert)

THE HIGHLANDS: AN IMPOVERISHED FLORA

Vast grasslands cover the central highlands, often known as the "High Plateaux", whose undulating hills lie at 1000 – 1500 m altitude. The grasslands are sometimes called pseudosteppes because of their sparse "bozaka" or bunch grasses: the Graminaceae *Aristida similis* and *A. rufescens* and some Cyperaceae. Here and there a few dicotyledons grow, either annuals or else perennials with a large root system that help them survive fire. A few lovely flowers catch the eye, such as the white blossoms of *Tachyadenus longifolius* (Gentianaceae), *Catharanthus lanceus*, one of the Madagascar periwinkles, the pretty wild *Gladiolus*, and *Clematopsis* flowers like the anemones of Europe, whose natural vine-like growth is reduced by the impact of fire to underground stems which throw up new shoots.

Other species which would normally grow as vines have similar adaptations to fire. These "pyromorphoses" are a witness to a recent past, when more or less continuous forest covered the present bare hillsides or *"tanety"*. One can still find vestiges of this forest near Tananarive, in ravines where a few trees and tree-ferns still grow. You can also find them in the form of sacred groves on the hills crowned by "rova", ancient princely strongholds, as well as regrowth in the old defensive ditches of hill-top villages.

These forests in general are low, with thickets, and a discontinuous herbaceous ground-cover. H. Humbert pointed out two species as characteristic: *Tambourissa* and *Weinmannia*. They are both trees of Asian affinity, with many different species that occur down to the east coast. The young leaves of *Weinmannia* tint the forest red and rose. Let us point out among so many others, the genus *Phyllarthron* whose species bear pretty, but quickly fading, purple and pink flowers. Children amuse themselves by tracing out words which remain indelible on the small leaves that grow on jointed, winged stalks — the last Merina queen supposedly left such messages.

Further north and northwest of Tananarive lie vast, cold, desolate highlands called "tampoketsa". Here, there are more numerous remnants of forest, sometimes even real woods, dominated by strange colonial *Pandanus*. These remnants as well as the widespread use of wood in houses and cattle corrals again show that forest was more widespread in recent times. Man has accelerated the natural deforestation which resulted from drier climate.

Here and there you find real moors with heathers (*Philippia*) and other species bearing small, often pungent leaves. The "vofotsy" (*Aphloia theiformis*) makes a tea dear to the Malagasy people. This is a little tree which varies in shape from one region to another, hardly recognizeable in Reunion, and whose range extends to the African coast. The "vofotsy" is known for its medicinal values, particularly as a slightly tonic diuretic.

In the final decade of the last century, man introduced *Eucalyptus*, particularly *E. robusta* and *E. rostrata*. They are now the commonest trees of the highland landscape, and furnish firewood, charcoal, housetimbers and even furniture. They have the great virtue of resisting fire, unlike more vulnerable pines. *Pinus patula*, of Mexican origin, forms 40,000 – 50,000 ha of forest plantations. There have been plans for huge pine forests north of Tananarive.

On the highlands, particularly in the regions of Ambatolampy and Antsirabe a "mimosa" species (*Acacia dealbata*) also dominates the landscape. It spontaneously invades uncultivated land, resists fire, and is another essential source of firewood and charcoal. In August, at the end of austral winter, the mimosa woods are a constellation of tiny yellow flowers.

The mushroom-lover can hunt through the moors and the Eucalyptus and mimosa woods in search of savory little mushrooms like red chanterelles. The walker can explore the rocky outcrops and domed granite hills of the plateaux, not so rich as those of the western slopes which we shall discuss later, but still with some fine plants: *Aloe* with fleshy leaves and handsome flowers, *Xerophyta*, strange little vegetables with narrow pointed leaves whose bases resemble a trunk, *Kalanchoe* with diverse leaves and flowers, and finally the beautiful *Angraeceum sororium*, an orchid with huge white flowers, which grows abundantly on the rock of Angavokely.

Around Tananarive in the Merina and Betsileo country, man has transformed the valley bottoms into rice terraces, at the cost of immense and meticulous labour. Natural vegetation or untouched swamps are now very rare. Still, in the flooded meadows near the little mud dikes of Betsimitratra, or by the inflow canals and on the terrace slopes in Betsileo country, we can find a few wild temperate genera: *Viola*, *Ranunculus*, *Epilobium*, and so on. These same genera, and sometimes the same species such as *Viola abyssinica*, occur on the mountains of East Africa. A curious little umbellifera, *Centella asiatica*, is frequent in wet places. It has round leaves, "frog-plates" to translate the Malagasy name "vilian-tsahona". It helps scar tissue to form over cuts and wounds, and is used both by local healers and European pharmaceutical companies.

Natural vegetation has little to offer on the highlands, thanks to human influence. But what a variety of cultivated crops appear in the Tananarive market! Just to list those cultivated in the immediate neighbourhood, there are tropical vegetables like sweet potatoes, manioc and taro beside temperate ones like potatoes, cabbage, cauliflower, carrots, turnips, radishes, leeks, lettuce, asparagus and artichokes. Among the fruits, the peaches, prunes, mulberries, apples, strawberries, grapes, apricots, mangoes, pineapples, oranges, loquats, mandarins, papayas, Chinese persimmons, bananas, and quinces mingle with still other fruits from the coast. Finally roses, gladioli and daisies are sold all year round, while the changing seasons bring cosmos, chrysanthemums, dahlias, mimosas, marigolds, pansies and many others.

How to describe the gardens, parks and avenues of Tananarive? There are plants of every country: guava and peach, Amazonian hazel and North American poplar. The residents of the capital have a taste for flowers: balconies and windowsills bloom with orchids, begonias and kalanchoes. The whole city turns red when the "Madagascars" bloom — *Poinsettia pulcherima* from Mexico, now the floral symbol of the island. "Madagascars" are two or three metre poinsettia bushes covered in flowers, (not the stunted Christmas pot-plants of the north). In austral spring the town turns to the bluish mauve peculiar to jacarandas (*Jacaranda mimosifolia*), another central American tree, mixed with the old gold of Australian *Grevillea robusta* or the pale violet of Persian lilac (*Melia azadirachta*).

Parc Tsimbazaza, created round two artificial lakes, assembles plants from the entire world alongside splendid Malagasy collections. You can admire the rock garden of succulents from Madagascar, Africa and America, strangely alike in spite of their different origins. The Botanic Laboratory of Tzimbazaza holds the national herbarium, a large collection of plants from all parts of the Island. There one can appreciate the immense labour undertaken since the forgotten navigators of the fifteenth and sixteenth centuries first made Malagasy plants known to the outside world, through recent scientists like H. Perrier de la Bathie, H. Humbert, and R. Capuron who have contributed so much to our understanding of the Malagasy flora. Before leaving Tananarive for regions richer in plants, we should make a rapid pilgrimage to the hillslope opposite Tsimbazaza. There, a few palms recall the first botanical garden of the island, and the memory of its creator the young H. Bojer, who died of malaria without ever returning to his native land.

WESTERN SLOPES OF THE HIGHLANDS: "TAPIA" WOODS AND NATURAL ROCK GARDENS

The western and southern slopes of the highlands have a specific botanical character, due to their climate and the low human population density.

Ever since Perrier de la Bathie, botanists have distinguished this region from the rest of the highlands. It is essentially a herbaceous landscape, a sort of fire-swept savannah, with interesting if small forest relicts, and scattered rocky massifs. Handsome granite monoliths stretch out to the west, on the road to Morondava, and to the south towards Tulear and Fort Dauphin. These border regions also include the magnificent sandstone fortresses of the Isalo, a bastion isolated from the rest of the highlands, the Bongolava or long mountain, the Anavelona, the Montagne d'Ambre, and the strange quartzite massifs of the Ibity and the Itremo, which also share some characteristics of montane flora.

The western slopes are almost entirely covered by savannah with *Loudetia simplex*, subsp. *stipoides*, endemic to Madagascar. This plant is gradually taking over the western savannahs, which themselves result from fires set by man. The original forests have left only scattered remnants near Tananarive, at Behenjy and Ambatofinandratana, and in the Isalo massif. This low forest, only 10 – 12 m high lets light penetrate and has a rich, varied undergrowth. The little trees have spherical crowns with twisted fissured trunks. Their leaves may be dark and dull, or, on the contrary, gleam gaily. You find few epiphytes here: ten or so orchids of the genera *Bulbophyllum*, *Angraeceum*, *Jumellea*, and few lianas.

The "tapia", *Uapaca bojeri*, dominates this woodland, the only one of the Malagasy species of this genus which is sun-loving and gregarious. It is also the most western species, for the others live in the eastern forest from the coast to the high mountaintops. "Tapia" fruits are edible, and sold as far away as the capital. This tree feeds the Malagasy silkworm, the "landibe" or *Boroceras madagascariensis*. Mingled with the "tapia" grow various sorts of Sarcolaenaceae, the largest of Madagascar's endemic plant families with ten genera and about thirty species. Perhaps we should mention in passing the other endemic families: Didiereaceae, 4 genera, 11 – 12 species, Rhopalocarpaceae, 2 genera, 14 species, Didymeleaceae, 1 genus, 2 species, Humbertiaceae, 1 genus, 1 species, Diegodendraceae, 1 genus, 1 species, and probably Geosiridaceae, 1 genus, 1 species.

The Sarcolenaceae, or Chlenaceae, are trees, bushes and shrubs with vivid flowers varying from red to yellow. Each massif of the western slopes seems to have its own Sarcolenacea, a good example of the extreme differentiation of living things in Madagascar. Among the "tapia wood" trees we may also point out *Asteropeia*, the only Malagasy representative of the Theaceae family. Like the "tapia" and the Sarcolenaceae, *Asteropeia* species are different from one locality to another.

These formations are relatively poor and often neglected by naturalists. Nevertheless they contain various plant species: Cunoniaceae of the genus *Weinmannia* which we have already met in the highlands, Anacardiaceae, Composaceae, and so forth. There is *Agauria salicifolia*, an Ericacea with beautiful bell-shaped flowers of white, pink or greenish hue, and several species of *Vaccinium*, some of which give pleasant fruits. These Malagasy "bilberries" are worth gathering, or even introducing into cultivation. In the Ericaceae, *Philippia*, the true heathers, have exploded into a considerable number of forms.

Floristically, the western slopes belong to the oriental domain, or windward flora, and form its limit to the west. Beyond that, on the west side, proper eastern groups become rare and African affinities predominate. Before leaving the western slopes, though, we must consider the rock plants. Madagascar possesses one of the richest xerophilous floras of the world. Life on rocks imposes harsh conditions of life: wide variation of temperature and humidity, which runs from total flooding in cups and hollows through to intense dryness. Roots creep into the smallest fissures, and little by little the plants themselves accumulate organic matter to create islands of vegetation on the hostile substrate starting with lichens and mosses which are the first colonizers.

These, plants are never ordinary, for they need extreme adaptations to survive. The commonest is to accumulate water in the tissues: the phenomenon of succulence. When the branches or twigs accumulate water as in the euphorbs, the resulting forms resemble American cacti. True cacti, or, better, the family Cactaceae, only come from America, except for a single species (*Rhipsalis cassytha*) which has reached Africa, Madagascar, and Ceylon. It is generally an epiphyte, but in Madagascar it takes it on itself to vary, and one very pretty form with red stems and fruits lives on the rocks. It hardly fits the stereotyped idea of a cactus with its cylindrical, thornless stems several millemetres in diameter, that grow in trailing tufts.

The succulent euphorbs are easy to recognize by their groups of two thorns growing either side of a small pad, the scar of a leaf petiole, as well as by their white latex which is corrosive and toxic. Be careful not to get latex in your eyes: at best you will suffer painful irritation of the cornea, at worst a conjunctivitis which can result in total blindness! Unlike the euphorbs, Cactaceae contain only a colourless and inoffensive liquid. *Euphorbia millii*, covered with spines, is sold commercially under the name "Crown of Thorns" or "Christ's Crown". Another group are called coraliform, for their branches like staghorn corals. To see extreme adaptations, look at *Euphorbia quartziticola* of the quartzite massifs of the Itremo and the Ibity, as well as *E. primulaefolia* and *E. moratii*, whose aerial portions are reduced to a rosette of leaves from which emerge coloured flower cups. The whole vegetative system is condensed to a large subterranean organ.

The genus *Pachypodium* (Apocynaceae), shows the same reduction. The walker often overlooks the "stone-plant", *Pachypodium brevicaule* of the Itremo and Ibity massifs. Trunk and branches are lost in a fleshy, swollen, pillow-shaped mass. You are suddenly surprised by yellow flowers on long stalks apparently springing straight from the earth, then, leaning down, you perceive the grey, stone-coloured vegetative mass that moulds itself into the surrounding rock.

Other yellow-flowered species of the same group have slightly more distinct above-ground parts, though still squat and fleshy. Their name means "thick foot", but in English they are sometimes called the "elephant foot". In Madagascar they may be called "dwarf baobabs". Some species reach a metre in height. They are very fragile, and usually perish if moved and yet far too many leave Tananarive airport for Europe. Let us respect them, let them remain where they are, products of years of growth in harsh surroundings — at most take home a few seeds to plant.

Another type of succulence may only affect the leaves, for instance in the genus *Senecio* which has so many species and varieties worldwide. The leaf may thicken without changing form (*S. melastomaefolius*), it may be transformed to a more or less furrowed cylinder (*S. sakamaliensis*) or the leaf may be so flattened laterally as to seem vertical (*S. crassissimus* and *S. cedrorum*). The vegetative parts of *Senecio* strongly resemble *Kalanchoe*. *Senecio*, however, has small yellow composite inflorescences, while *Kalanchoe* has delicately-coloured flowers like little bells.

Kalanchoe has some sixty species in Madagascar. Like virtually all the genera which grow on the rocks it occurs mainly in the west and south, not the humid eastern forests, barring a few gorgeous exceptions. This genus possesses vast power of vegetative reproduction, by leaf buds, flower buds, and stolons. The plant lover may safely propagate kalanchoe, which should grow to his complete satisfaction. It is impossible to list the *Kalanchoe*, but let us point out *K. synsepala* of amazing variability, scarcely the same from one rock to the next, and *K. pubescens*, supposed to bring wealth to whoever grows it. We can try the experiment: if not fortune, we shall have the pleasure of growing this lovely plant!

Other plants grow on the rocks, whose splendid red or orange flowers blaze in the landscape: about 50 odd species of the Aloe. Aloes are common in the south and west, but only grow in rocky areas on the highlands. They have the familiar characteristics of great variability from one locale to another, and very limited distributions of each form, sometimes confined to a single rocky dome. Here are two examples: *A. compressa* of the Ibity quartzites has one variety, *rugosquamosa*, in the quartzites of the Andrantsay basin near Betafo, and another variety, *schistophila* in the Itremo schists. *A. capitata* has three morphologically distinct varieties: *gneissicola*, *quartzicola*, and *ciponicola* which inhabit as their names suggest, gneiss, quartz, and marble. *A. capitata* is one of the crowning jewels of Malagasy nature. A group of plants sends up 20 to 30 flower stalks topped by globular inflorescences. The scarlet of unopened flowers at the centre of the stalk harmonizes with the sulphur yellow of open flowers at the periphery. *A. c. ciponicola* are legendary for routing Sakalava troops who threatened to conquer the plateaux: in the morning mist the aloes seemed like an army of soldiers bristling with arms, ready to repel the invaders!

We must also mention the strange Labiacea which put out short twigs that bear small, fleshy, triangular leaves in the dry season, and then in the rainy season large normal leaves.

A second way to resist dryness is to reduce the leaf surface, thus reducing transpiration. On rocky

Fig. 2.3 *Aloe capitata* on the granite outcrop of Angavokely, nr. Antananarivo (A. Jolly)

outcrops we may see, in southern springtime, the pretty violet mauve corollas of *Xerophyta* (whose evocative Greek name means plant of dryness). They are upright plants which mimic little bushes, constructing a sort of trunk out of leaf sheaths tightly bound together, and tunnelled through by adventive roots. Here again there are a wealth of local forms. Some *Xerophyta*, like *X. daslirioides*, can reach considerable age and nearly two metres in height.

The Papilionaceous *Mundulea phylloxylon*, of the Isalo massif, seems to have lost its leaves and flattened its twigs to resist dryness. We would think so, except that another genus of this family has the same structure in the humid forest of the east: genus *Phylloxylon* with the same name as the species of the Isalo. *Mundulea* looks like a grey green broom, with pretty little pale yellow flowers at the start of the rains.

Plants can also combat dryness by simply letting themselves dehydrate for long periods. Several Pteridophytes do so, and oddly, so does a flowering plant (*Myrothamnus moschatus*) of the family Myrothamnaceae, represented by just one species of the same genus outside Madagascar in South and East Africa. All these plants dry out, seem dead, break at the slightest touch, but grow green again in a few hours of moist atmosphere!

Many other plants people the rocks: Graminaceae and Cyperaceae, Orchids including Cynorkis which raise their spikes in the midst of grassy fields at the start of the rainy season, or the large *Angraceum sororium* and *Sobennikoffia humbertiana*, Commelinaceae, Asclepiadaceae, and Labiaceae like the amazing *Perrierastrum oreophilum*, a bush with fleshy, serrate leaves.

We could almost claim that every rocky hill holds a unique botanical community. Every climb brings a new marvel or botanical enchantment — still, one must remember that such riches must be preserved: even with the best intentions, never uproot such plants to bring them home. It is always preferable to collect seeds, which does not endanger the wild population and helps us understand the plant better through its development.

Before leaving the highlands and their slopes, and in spite of the temptation to descend to the west raised by the rock plants with their evident western affinities, we shall climb to the highest Malagasy summits.

THE MOUNTAINS: HEATH, LICHEN WOODLAND, MOSS FOREST

From north to south rise the mountain massifs of the Tsaratanana, (highest point of the island at 2876 m), the Marojejy, the Ankaratra, the Andringita and the Andohahela. The lower massifs of the Ibity and the Itremo, already described, belong to the same Malagasy montane category.

The entire summit vegetation of the Tsaratanana was destroyed at the beginning of the century by the geologist Lemoine, of the Museum d'Histoire Naturelle de Paris, who enthusiastically described the fire ravaging the mountain! In 1912 H. Perrier de la Bathie, a geologist himself but also one of the fathers of Malagasy botany, found a few strips of forest which had escaped the disaster. On his later expeditions of 1923 and 1924, there was none left. In the interval, other missions directed by ignorant men had again laid fires! We shall never know what we have lost irredemably in these flames, fruit of pure human stupidity.

The Marojejy is relatively near the capital but more recently explored by botanists. H. Humbert, R. Capuron and G. Cours-Darne explored it in 1948 – 49, only eleven years after the first ascension by a European. It remains intact in all the splendour of virgin vegetation. When one has the privilege of admiring the small summit of the Marojejy (2135 m), this "natural marvel" which H. Humbert revealed, one regrets all the more bitterly the ruin of the Tsaratanama.

The Ankaratra near Tananarive, dominating the pretty village of Ambatolampy, has only minor botanical interest. The main peak, at 2643 m is Tsiafajavona, "He whom the clouds never leave", but how many times has it burned? Herds pasture on the slopes and the herdsmen set fires for the "green bite", a growth of new shoots in the dry season. The traveller may visit the forest station and easily reach the summit, and even suffer biting cold when wind and rain arrive together. But the original vegetation is mostly replaced by an immense reforestation of pines.

The Andringitra whose summit, Pic Boby, reaches 2658 m, is easy to see but not so easy to climb. The trek up to the Andoharina plateau at 2000 m demands considerable endurance. There, one can admire marvels of the vegetable world like *Panicum cupressoides*, a little Graminacea which mimics cyprus branches. It forms handsome groups in humid places like shallow depressions or along water-courses. There you also find a Halorrhagacea, *Gunnera perpensa*, which is certainly of austral origin. The same species exists on the Abyssinian highlands in similar environments. The Andohariana plateau also offers us fine representatives of temperate genera (*Rubus*, *Ranunculus*, *Geranium*, *Alchemilla*), and few large Ombellifereae, which likewise grow in East Africa. The strange affinities and strange distributions of these mountain species, are very difficult to explain. One can imagine a double colonization, by the north temperate species listed above, and by southern species including *Gunnera*, *Kniphofia*, and *Stoebe*, extraordinary compositae with shape and leaves like heathers, as well as *Philippia*, the true heathers, which have a few species in the Mascarenes and East Africa, but prodigious speciation in Madagascar.

This tiny plateau contains several high level endemics!

The ascent of the summits of the Andringitra is not possible for everyone, but what joy to climb these naked, vertiginous rocks. You reach the Pic Boby by the eastern face. It took H. Perrier de la Bathie and J. Descarpentries, a topographic surveyor, three attempts in 1922. They say their dog Boby was first to the summit . . . and gave it his name: even geography can sometimes allow a little humour. On these peaks where the temperature can fall to – 15°C, snow has been seen three times in twenty-five years.

The final massif, the least known, is the most difficult to reach: the Andohahela. About 50 km north of Fort Dauphin, it rises to 1956 m. It lies between east and western regions, one side rainforest, the other dry savanna, and so it is a striking microcosm of the whole island.

H. Perrier de la Bathie gave concise, perceptive names to the three main mountain communities: "moss forest with herbaceous undergrowth", "lichen woodland" and "montane heath". The physionomy and unique quality of the mountains seem summed up in a few words. If one must integrate these communities into a more general classification, we can use the banal descriptions: dense humid montane forest, sclerophyllous forest, and montane bush. This makes comparison easier and definitions more exact, but how much less charming! We shall find the charm again in nature itself.

The "moss forest with herbaceous undergrowth" is only easily visited on the flanks of the Ankaratra, but it is already impoverished there by the herds. Elsewhere it demands real expeditions. This plant community is scarcely 15 m tall. The treetrunks twist, with many branches, covered with mosses and liverworts whose leaves trail in draperies of old gold, greenish red, or nearly black. Epiphytes abound, using other plants for support not nourishment. Orchids of varied shape and colour, and delicate mountain Kalanchoe would be well worth cultivating. There are amazing *Senecio francoisii*, with somewhat succulent leaves brightly tinted with violet on the underside, and *Medinilla*, (Melastomataceae), little bushes with delicate mother-of-pearl pink flowers. There are innumerable ferns of all shapes and sizes, *Peperomia* of which some might be cultivated like foreign ones of the same genus, *Rhodocodon* with delicate flower spikes like lily of the valley. A whole strange, magnificent world of flowers glides over trunks, branches, rocks and soil in perpetual oozing humidity.

The trees generally have small, thick leaves (sclerophyllia) and belong to several families represented at lower altitude. There is one exception, the Aquifoliaceae to which belongs the European holly and the South American holly whose leaves are brewed to make the tea called mate. There is only one representative in Madagascar, *Ilex mitis*, which also grows in East Africa. In the montane forest nearly all the species are unique, having slowly evolved out of the communities of low altitude. There we find the only Malagasy Gymnosperm genus, *Podocarpus*, with several species from the mountains down to the east coast plain. *Sloania* grows there with big, decorative yellow and red flowers, and several ornamental *Grewia* and *Symphonia* species. One single *Symphonia* exists outside Madagascar, *S. globulifera*, the same species in Africa and South America which is extraordinary enough. In Madagascar we are not yet sure how many species there are: a real pulverization of forms, again ranging from the East Coast to the highest mountains. The mountain species are particularly decorative because they grow in horizontal layers, leaves as well as flowers. One species with bright red button-like flowers is sold before Christmas for house trimming.

The bamboos, those giant grasses, have one species in each massif. The tree ferns, two genera of Cyathaceae, have more than forty species. We frequently find tree ferns near torrents, their trunks surmounted with a parasol of fine lace fronds raised 10 – 15 m high toward the light. Sadly they are over exploited, their trunks cut up for flowerpots, the "*fangona*". Thus, these marvels of the vegetable world may disappear, though they merit rigorous protection.

The undergrowth is burgeoning with soft plants: ferns, many coloured *Impatiens*, delicate *Streptocarpus*, innumerable *Begonia*. Anyone who knows something of European plants will be surprised to reconize familiar genera or even species: *Sanicula europaea*, *Viola abyssinica*, *Caucalis melanantha*, *Cardamine africana*, *Alchemilla*. We can wonder what the peregrinations of these temperate plants, which brought

Fig. 2.4 Montane moss forest, on the Ankaratra Massif. An epiphytic euphorbia blooms on a headless treefern trunk in the foreground; mosses coat the branches of the tree behind the sunlit treefern (A. Jolly)

them to Madagascar down the backbone of East Africa.

Above the humid forest, climatic conditions grow harsher. The temperature range and intense sunlight which can lead to extreme, if temporary dryness, and here the vegetation is even more markedly adapted. In the woodland with lichens, leaves are reduced; none exceeds 20 cm length except for a few palms and *Pandanus*. There is a far greater richness of species than in the more humid forest below. On the ground the foot sinks in a continuous carpet of splendidly coloured mosses. This carpet seems to climb the trunks and branches, hanging from the twigs in fine lace. On the highest branches, long beards of *Usnea* float in the wind. All this leaves little place for herbs on the soil and for epiphytes on the branches — still numerous but less than in the humid forest.

As conditions become ever more severe, upward to total absence of soil as on the Andrigitra, the vegetation changes to montane heath and then to rock formations very different from those at lower altitude. Montane heath forms a dense thicket where all the plants look like heathers. One genus, *Erica* are the true heathers which give the name ericoid, but plants of many groups grow in similar manner: Composaceae (*Psiadia*, *Helichrysum*, *Stoebe*, *Stenocline*), Ericaceae (*Philippia*, *Agauria*), Podocarpaceae

Fig. 2.5 *Pachypodium rutenbergianum* flowers near Diego Surarez (Ph. Oberlé)

(*Podocarpus*), Rhamnaceae (*Phylica*), Thymeleaceae (*Gnidia*), Melastomaceae (*Osbeckia*) and so forth. Even the bamboos reduce their leaves!

"Lichen woodland" and "ericoid heath" rival each other in splendid forms and colours, and by the grandiose high mountain landscapes where mists nearly always float, making the smallest sunbeam explode with beauty. These communities alas, are extremely fragile. Only the summit of the Marojejy remains inviolate, and one dare not imagine the disaster which would follow the slightest imprudence there.

The rock plants, of course, are less threatened because they cannot support continuous fire. Those the Andringitra count among the most interesting. We find there fantastic endemism, with, among others, *Helichrysum stilpnocephalum*, the only representative of an East African group, and *Sedum madagascariense*, the only Malagasy representative of its genus. We cannot mention them all! Let us insist, instead, on the originality and diversity of the Malagasy mountain flora. Each massif has its own species, usually local representatives of widespread complexes, but sometimes completely isolated. This diversity of speciation is common to many regions of the Great Island.

SAVANNAHS AND DRY FORESTS OF THE WEST: KINGDOM OF BAOBABS

The occidental or "Leeward" domain is 80% herbaceous formations, real savannahs of large flat-leaved grasses, sometimes more than a metre tall. It has only scattered bushes and trees, and is regularly swept by fire. Peasants set the "bush fires" to clear before cultivation, and to burn the dry parts of the grasses so the roots will sprout new leaves in the dry season. We have already spoken of the importance of fire and its influence on vegetation and soil. Fire has many whole-hearted adversaries, and few who champion it without reservation. But men must live, and their cattle need grass. Therefore one should control fire, and set it at the beginning of the dry season when there is not much dead material, so the flames do not remain too long in one place, or seriously penetrate into the forests. These early, controlled fires, allied to rational cattle husbandry with balanced stocking levels and grazing rotation, could suit Madagascar as they do other cattle-raising countries.

Lovers of elegant grass species could make a fine harvest in the west. They could even benefit by one of the few Malagasy botanical books which is readily accessible, the recent Flora of the Graminaceae by J. Bosser, with its very beautiful illustrations by Edmond Razafindrakoto.

A few coloured plants and flowers enliven the prairies, which are called "bozoka". *Crotalaria*, with *Indigofera* with violet red flower spikes, a few terrestrial orchids, *Lissochilus* and *Cynorkis*, *Aloe contigua*, and the lovely *Crinum voyroni* with its white lily flowers. Let us also note a few graceful little *Dioscorea* plants, the genus of many edible tropical roots including the yam.

The trees, relics of the original forest, are small sized. They seem to have little means of protection from fire, unlike those of the East African savannahs. Instead, they survive thanks to their capacity to regenerate from suckers. Vines, as on the plateau, change shape and either half-bury themselves like *Aristolochia acuminata*, or turn into bushes like *Cryptostegia grandiflora* which gives a kind of rubber. (It has even been planted for that purpose.) The perfume of its huge mauve flower can kill the unwary or desperate soul who breathes it.

Tamarindus indica are abundant with their refreshing pulp, and *Pupartia caffra* with acid fruits smelling of terebenthine, a few pretty *Terminalia* whose branches spread in superimposed layers, *Acridocarpus excelsus* which varies from one site to another, and finally *Dicoma incana*, a surprising and beautiful arborescent composite. Even in an impoverished flora, we cannot cite everything. Only a word on the palms, *Medemia nobilis*, *Borassus madagascariensi* and *Hyphaene shatan*, which are endemic species of African genera, in contrast to the palms of the East which are wholly unique to Madagascar. The two first species probably originated in the gallery forests and spread from there into the surrounding savannah. *Hyphaene shatan*, whose stems divide several times and group in tufts, characterizes certain dry habitats.

In summary, these savannahs are poor in both vegetable and animal life. Aside from the zebu in their uncountable and omnipresent herds, there are no mammals, no birds. Their only interest lies in their origin. The poverty of this country is recent, and the traveller can still imagine what these vast, barren spaces were like before they were ravaged by fire. He can seek out in the forests which still persist here and there, a vision of the first men arriving in these regions scarcely more than a millenium ago, when the great birds and giant lemurs still lived.

These western forests were estimated in 1960 to cover more than 2,600,000 ha, whether or not degraded. They lie mainly at low altitude, near the coast. They extend from the south where they fringe the xerophile bush, all the way to Diego Suarez beyond the humid intrusion of the Sambirano.

They are essentially deciduous, that is, most trees lose their leaves in the dry season. Then they take on a sinister aspect. The only touches of colour are the arborescent euphorbias, the tufts of *Lissochilus*, and the magnificent blossoms of leafless vanillas.

There are great floristic and physionomic differences depending on the soil: clay, sand, or calcareous earths. The calcareous massifs offer most surprises and oddities. They have eroded into extraordinary "karst landscapes" in the Ankarana near Diego Suarez or the Kelifely, in the Antsingy of Namoroka or Antsalova. Their name "tsingy" is supposed to be the sound which a blade of limestone makes when you strike it. These massifs, cut through with gorges and caverns, sheltered brigands in the olden days and refugees in periods of insecurity. There is a mosaic of plant formations according to the degree and nature of the erosion, from little forests on flat, well drained blocks, to bare rock where a few pioneers cling. In spite of their difficulty of access (or rather thanks to it) the "tsingy" are a botanist's paradise. *Dalbergia*, the rosewood, dominates the forest, *Commiphora* with odorous resin and wood which was (and still is) used to start fires by friction, and *Hildegardia* with its lovely red flowers.

As usual in Madagascar, there is prodigious richness, originality, endemicity. Furthermore plants here grow into very strange shapes. Succulence takes different forms from those we have seen on the rocks. *Euphorbia enterophra*, a large tree with flattened green twigs or cladodes, presents, from afar, the silhouette of a Mediterranean parasol pine. Leafless lianas abound, for example the vanillas with white or yellow flowers more or less tinted with carmine and red. They are near relatives of the culinary vanilla from Mexico which is cultivated on the east coast. They do not produce aromatic fruit, but are said to be aphrodisiac.

Plants with succulent leaves are rare; a few *Kalanchoe*, *Aloe*, and the related genus *Lomatophyllyum* which is distinguished from *Aloe* by the fleshy, indehiscent fruits. Just one *Senecio*, *S. sakalavarum*, grows in these forests. There are several forms with swollen trunks, lianas or bottle trees of diverse families. Among the lianas in the family Vitaceae, (which includes grape vines), there are several *Cyphostemma* species whose conical trunk may grow 2 – 3 m in height on a base 40 – 50 cm in diameter, or with a gross semi-underground organ more than 1 m in diameter for 30 – 40 cm thickness, such as in *C. elephantopus*. Its spiral tendrils and stems only appear in the rainy season; in repose it looks like an old half buried tyre! The Passifloracea or passion fruit family includes bottle-lianas of the genus *Adenia*.

Pachypodium, which has dwarfed stone-plants on the plateaux here turns into thorny long-necked

Fig. 2.6 Africa has only one species of baobab; Madagascar boasts seven. *Adansonia grandidieri* in the wet season tower above the deciduous forest near Morondava (R. Albignac)

Fig. 2.7 Baobabs near Belo sur Tsiribihina in the dry season (Ph. Oberlé)

bottles. *P. lamerei* has groups of three spines, *P. rutenbergianum*, *P. meridionale* and *P. sofiense* groups of two. All three of the latter come from the north of the island, while the former grows south of Morondava. These plants can reach some 10 m in height.

The genus *Adansonia*, named for the great French naturalist Adanson, has an analogous appearance. Everyone knows the baobab (*A. digitata*) as a characteristic tree of Africa. Most people do not know that Africa has only one species, Australia has one or two, while Madagascar has seven clearly distinct ones.

The African species grows in Madagascar near Majunga, with a famous specimen in the town near the sea front. The seven endemic species have very different ranges. In the western forest, *A. grandidieri* is a splendour of the plant world. I cannot convey the beauty of a baobab grove, when the majestic trunks redden in the light of the setting sun. The seeds furnish oil, and men harvest them acrobatically, scaling the sheer vertical trunk by means of wooden spikes driven into the bark. The trunks are gorged with water. In the lean season, herdsmen cut down a few trees, and open them to feed the herds. It is scarcely nourishing, and you can count the ribs of the cattle fed thus, but they can survive on the watery pulp until the new grass grows. The trunks are sometimes fashioned into real cisterns which collect water running from the top of the tree. Among all the other uses, *Adansonia* bark can be used for making rope and cattle-leads.

Let us mention some of the other bottle trees: *Moringa drouhardii*, *Givotia madagascariensis*, *Gymnocarpus americanus*, before passing on to other adaptations against dehydration such as leaf reduction, sclerophyllia, and drying then reviving. This last trait is spectacular in the pretty fern *Platycerium quadridichotomum*, the only species of its genus to leave the humid forest, doubtless because of its potential ability to dry and revive. Bottle vines, as we have seen, and also *Begonia* resist drought by losing all or most of their aerial parts. The only two western bamboos lose their leaves in the dry season.

The Malagasy banana (*Ensete perrieri*) of the western forests completely loses its leaves, and is reduced to a pseudo-trunk formed of leaf-sheaths. The genus *Ensete* is best known for its Ethiopian species which forms the staple diet for several human populations of the south Abyssinian Plateau. It grows in drier and colder conditions than *Musa*, the true bananas, but still it is amazing to find such close coordination between the environment and the plant's reaction.

We cannot mention all the plants of the west. There are too many known, and too many more not even catalogued, let alone described in their organization, intimate physiology, ecological needs. Will they be known one day? A few have reached European knowledge, like the strangely tortured canes made of "manjaka betany", the "Great King of the Earth". Traditionally reserved for great chiefs and eminent dignitaries, the canes are now sold to tourists. They are cut from the trunks of young *Baudouinia rouxevillei*, a legume which would be quite banal if its growth did not show such strange anarchy. We may also point out the genus *Uncarina*, which also has several species in the extreme south. Its fruits have long rigid arms, ending in recurved hooks. If you pick one up it sinks in your fingers, to be removed

Fig. 2.8 *Adansonia za* grows only in the Mandrare valley, and may become extremely large (A. Jolly)

Fig. 2.9 *Adansonia fony* grows on the dry Mahafaly plateau. This is the dwarf baobab, only 2 – 5 m high, perhaps the most grotesque of all (H. Rabesandratana)

with difficulty. Peasants use them as traps: they put bait in the middle of a pile of these burrs, and the rat or mouse which enters does not escape! The genus was first called *Harpagophytum*, the harpoon plant, but nomenclatural rules have replaced it by *Uncarina*. The same genus occurs in South Africa, showing once again the strange affinities across the Indian Ocean. The same is true of the genus *Delonix*, with the flamboyant *D. regia*. This magnificent tree which ornaments so many tropical countries originates from the Malagasy *tsingy*. J. Leandri, a botanist from the Museum national d'Histoire Naturelle de Paris, discovered it there in 1933, solving the much-debated mystery of its origin. Several other species of the same genus merit cultivation, and also its near relative *Colvillea racemosa*, whose colour outshines the flamboyant, but which has scarcely left its native land where it borders the streets of a few small southern towns.

THE SOUTHERN BUSH, A UNIQUE HABITAT

The southern bush is the most specialized of all Malagasy habitats. About 48% of its plant genera and 95% of its species are endemic to Madagascar.

The flora of the southern "bush" is so original in both its endemicity and its adaptations that it will be described in detail in the following chapter.

LUXURIANT FOREST OF THE EAST

We shall climb to the eastern forest by the col of Ranopiso, near Fort Dauphin. There we can travel within a few kilometres from the subarid vegetation of the south into the great rainforest, across all the vegetation types of the island. In R. Battistinis' lovely phrase, this is a "rainfall fault-line", an abrupt transition from 600 mm to 1800 mm of rain per year.

We are here, then in the rainforest, so much described, so much desired, so much laid waste. Man used to live in equilibrium with this environment, growing dry-land or "mountain" rice which the first immigrants brought from Asia a thousand years ago. In that traditional form of agriculture, the rice was sown in clearings where trees were felled and burned within the forest. After one or two harvests the field was abandoned to forest regrowth. This kind of agriculture is universal in tropical countries wherever forest grows. It is called "tavy" in Madagascar, swidden or slash-and-burn, and sometimes "essartage" when describing the same practice in neolithic or mediaeval Europe. Outsiders have been shocked, and have accused the peasants who practice "tavy" of practically every evil. However, this type of agriculture is well adapted to the environment, for farmers with limited technical means. Fertilization by ash, fallow with forest regrowth, and the forest in turn rebuilding the soil is a simple and judicious use of natural processes. This system of traditional agriculture gives maximal yield for minimal human effort. Only when the human effort is subsidized by dead forests of the coal age, in modern oil-based fertilizers and herbicides, and oil-run machinery, can we achieve more yield per man hour than the peasant who uses live forest as his subsidy.

Of course, the peasant does lay waste the forest for small return. However, every present-day use is wasteful. Either we clear-fell, burn and plant permanent crops, at most saving a few of the more valuable timber species, or else we extract a few cubic metres of logs per hectare with an armament of tractors and machine saws and fuel oil.

"Tavy" agriculture at least has the merit of only using the forest temporarily. We certainly do not advise the spread of such agriculture in the future; there must be new methods to come, but let us be wary of simply copying the methods suited to other ecological regions. The tropical rainforest is a unique habitat; let us seek unique means of using it. The people of the forest have much to teach scientists and agronomists. Let us be humble, and listen to the voice of their ancestral wisdom.

It remains true that the tropical forest is in danger and needs protection against all sorts of human appetites. Fortunately in Madagascar steep mountainous topography discourages both commercial platantions and industrial logging on the eastern slopes, outside the coastal strip. Beside the aesthetic, ethical, and ecological aspects of forest preservation, we must underline its role in protecting the climate and the watersheds, and the latent genetic potential of this complex ecosystem.

H. Perrier de la Bathie censused 100 m² of forest in the Maroansetra region; he found 102 plant species belonging to 35 families. When he counted 100 plants growing side by side he found 53 species of 24 families! Can we even imagine the possible uses of all these organisms? In the 100 m² there were 6 Apocynaceae, one of the most important medicinal plant families, with high alkaloid content. The rainforest species generally do not survive the destruction of the forest. They are doomed to disappear, or at best to lose a large part of their variability through habitat restriction.

Traditional Malagasy herbalists use a large number of plants of supposed curative power. Many of these remain little known to science, and further studies could reveal the value of this immense natural treasure.

Wild Malagasy coffees are now found almost only in nature reserves. We already know that in Madagascar these plants contain only tiny amounts of caffein. It would obviously be profitable to grow them either as is, or in the form of hybrids with African species. In the only reserve of lowland rainforest, we found 5 different coffee species in 2 km. They only exist because the forest is protected. This type of lowland forest has practically disappeared after many years of overexploitation, except for this one

Fig. 2.10 Lowland eastern rainforest rises in tiers on the island reserve of Nosy Mangabe in the Bay of Antongil (Ph. Oberlé)

Fig. 2.11 Rainforest at middle attitude or further south grows to smaller stature. This humid forest, in the reserve of Andohaleha lies below the Tropic of Capricorn. A slash and burn clearing left the grassland and dead tree in the foreground (A. Jolly)

reserve, and for a fine but unprotected massif north of Fort Dauphin where you can also admire the endemic family, the Humbertiaceae. (It is not dedicated to the great botanist Henri Humbert, but to an earlier homonyme.) This localized but abundant species is highly resistant to decay and fire. Its wood is sometimes used for firewood, or to produce an odour like sandalwood. Its stripped fire marked trunks stand isolated in the savannahs, mute witness to forests that have vanished.

The forest of the east is not hostile, in spite of the descriptions of the first foreign travellers, or the fear which afflicts the plateau peasants in their rice fields. For those who live there, the forest furnishes animal and vegetable food, remedies, shelter, tools, weapons, clothing and amusement.

Let us follow a Betsimisaraka peasant, wearing his shift of raffia fibre (*Raphia ruffia*) and his hat woven of Cyperaceae palm-fronds. Over his shoulder he carries his "angady", a long hoe hafted with *Phylloxylon* or some other hardwood. As he walks to his rice or manioc patch he gathers wild yams, young shoots of *Dracaena*, and fronds of *Marsillea* and *Stenochaena tenuifolia* which will enliven the dinner his wife will cook. Palm hearts form an elegant vegetable, though the bitter heart of the Ravenala is less appreciated. *Dilobeia thouarsii*, one of three Malagasy Protraceae with amazing leaf polymorphism, provides a vegetable oil. In famine periods he can make do with seeds of *Cycas thouarsii* or *Typhonodorum lindleyanum*.

His traditional house is made of wood, without the aid of metal parts. The walls are hewn from the outer trunk of a ravenala, flattened under weights after cutting. In a continuously clement climate this is enough shelter. Inside a woven mat of *Pandanus* leaves lies on the plank floor, which is raised on stilts above rain-water, mud, and chickens. Recipients of wood and bamboo hold the household food and water. This brief account only tells a tiny part of his multiple uses of the forest plants.

But let us now penetrate into the heart of the forest. The canopy, on average, reaches no higher than 30 m except a few emergents like *Sloanea rhodantha* and the "ramy", *Canarium madagascariense*, whose wood is used for light, veneered furniture. We are surprised by the floristic and biological diversity, complexity of the plant world equalled in no other habitat. Furthermore, the eastern forest and its western extension the Sambirano has 90% of its species endemic to Madagascar.

Many of its taxa are highly localized. In all we find about 40 genera and more than 1200 species. Many families and genera are mainly confined to the east, like the Palms with 19 genera and 130 species, almost all of them in the east, as well as the Myristicaceae and Trichopodiaceae species, and the handsome genus *Pandanus*.

Let us linger a moment with palms and pandanus as examples. Of the 19 genera of palms, only 7 are not endemic — each of these contain only one species. Except for *Raphia ruffia* all these grow in the East but have close relatives in Africa. The 12 other genera, with more than 120 species, are endemic but with Asiatic affinities, and the great majority localized in the East. Only a few species of Chrysalidocarpus and Ravenea penetrate the Western Domain.

About 30 of the 70 *Pandanus* species live in the east. This genus includes more than 700 Pacific species, but few in Africa. The *Pandanus* in Madagascar have colonized the high mountains of the far south and the western "tsingy", whereas in the East they proliferate in swamps and along the coastal forest. Among the many diverse forms, *Acanthostyla* species are most spectacular: living obelisks, with a little plume of stiff leaves at the top of a tall mast, which in turn emerges from a cylinder of short sprouts.

Palms and *Pandanus* grow throughout the Indian Ocean, but every island has its endemics. The only species common to them all is the coconut. Mauritius and Reunion only share one of their palms and one Pandanus. The extraordinary "coco de mer" (*Lodoicea seychellanus*) is unique to the Seychelles, with the largest seed in the plant kingdom. Palms and Pandanus reveal the lack of unity in the flora of the Indian Ocean, and would well repay further study by biogeographers.

The biological adaptations of the rainforest resemble those of the rest of the world's rainforests. There are trees with stilt or buttress roots, though not so highly developed as on other continents. There are abundant lianas, cauliflory (flowers growing directly from the wood) is relatively frequent, and many epiphytes. Epiphytic ferns and Orchids grow in association, though we do not know why. *Ophioglossum pendulum* always grows with *Asplenium nidus*, and *Oleandra africana* is often with *Vittaria scolopendrina*. The beautiful orchid *Cymbidiella rhodochila* always accompanies the no less beautiful fern *Platycerium madagascariense*.

We list more than a thousand Orchid species from Madagascar. All are pretty, but most small and discreet. A few stand out for form and colour. *Cymbidiella rhodochila* has flowers in bunches 10 cm in diameter, greenish yellow, leopard spotted with large green dots. *Aeranthus* species have hanging inflorescences which can reach 60 cm long, with large, very long-spurred green or white flowers. *Eulophiella roempleriana* seems to live exclusively on *Pandanus* in marshy regions. Its metre-long flower stalk carries leaves of the same length. There are 15 – 25 red flowers on the stalk, each 8 – 9 cm in diameter.

Orchids are pollinated by birds or insects which come to lick the nectar. They pick up pollen grains and transport them to other flowers. They are sometimes so specialized that only one species of insect can fecundate the flower.

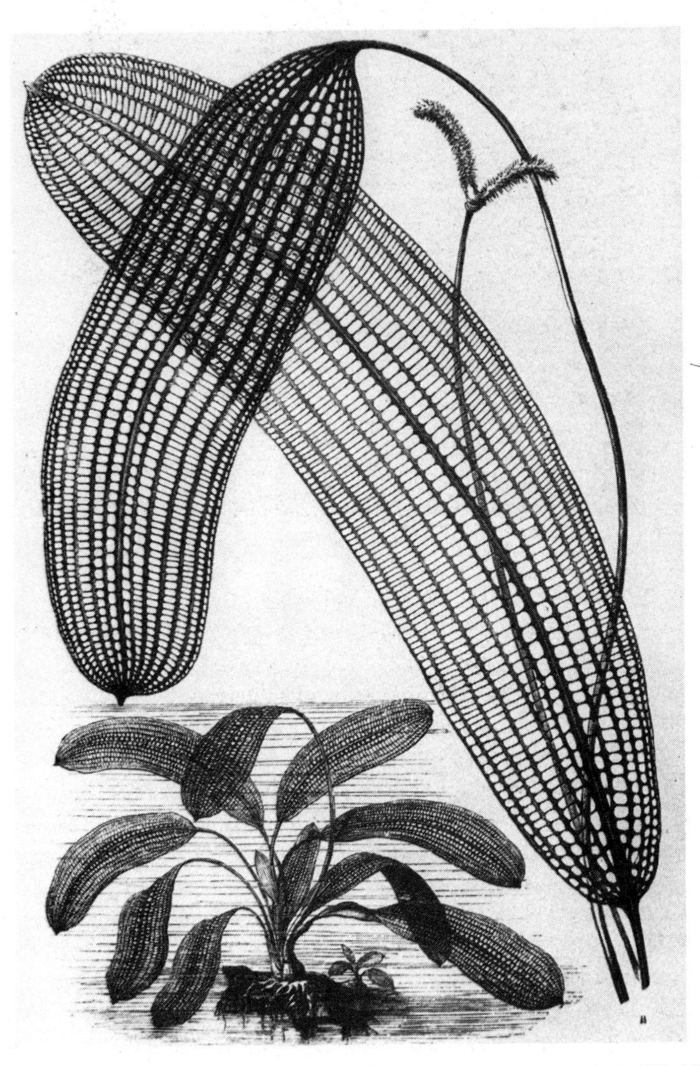

Fig. 2.12 *Aponogeton fenestralis* grows in streams of the eastern escarpment (W. Ellis 1858)

Fig. 2.13 Madagascar has more than 1000 orchid species, most in the eastern forest (A. Jolly)

The three white flowers of *Angraeceum sesquipedale*, most famous of Malagasy orchids, light up the green gloom of the forest like stars. The orchid bears a spur 35 cm long. Alfred Russell Wallace, studying this flower, concluded that a sphinx moth must exist with a tongue 35 cm long to reach the nectar in the spur. At the time entomologists laughed at the idea...but forty years later they discovered the moth and named it *Xanthopan morgani praedicta*, in honour of Wallace's prediction. Its proboscis unrolls to more than 30 cm, but only when it nears the orchid. Nature has foreseen everything. As the insect is nocturnal the flower gleams with a luminous whiteness that guides it through the night. Further, as the insect is fairly rare, the flower remains open and fresh for weeks, waiting...

Vanilla which also belongs to the orchid family, can only be fertilized by one bee species which lives in Mexico and Guyana, the edible vanilla's orginal home. The vanilla which forms one of Madagascar's chief exports, and the riches of the Antalaha region, must thus be fertilized by hand.

Among the vanillas nature offers us another example of the advantages we might gain by studying plants. In fact, Madagascar has several species of wild vanilla, which do not produce the perfumed pods of their Mexican relatives, but are far more disease resistant. In particular, they can resist the fungus infections which regularly devastate vanilla nurseries. It should be possible to create hybrids with the advantages of both strains.

There are many other lovely epiphytes in the forest: Melastomaceae of the *Medinilla* genus, *Kalanchoe*, *Clerodendrum* (Verbenaceae), at least two *Solanum*, and so on. To find them, either scan the treetops with binoculars, or go to tavy clearings and logging sites. Epiphytes fallen to the ground die slowly, so you can collect and try to grow them. On the other hand, it is inexcusable to gather such plants inconsiderately. Some species of Orchids, too much sought after, are already disappearing

Herbaceous plants are rare in lowland tropical rainforest. There are some Selaginella, often tinged with metallic blue, and fairly numerous plants whose leaves are geometrically spotted with white, pink or mauve, particularly among the Labiaceae, Acanthaceae, and Melastomaceae. There is a succulent, *Euphorbia lophogna*, rare Cyperaceae and bamboo-like Graminaceae, a sensitive plant, *Biophytum forsythii* of the Oxalidaceae which closes it's leaves at the slightest contact, a few ferns, and finally some palms so reduced in size they seem like herbs.

The Malagasy forest is remarkable for its flowers, which is rare in this type of vegetation. *Ardisia* for instance, bears large bunches of pink flowers, while *Symphonia*, *Sloanea*, *Tisonia* and *Dichaetanthera* flowers are red. Red and orange predominate. In the undergrowth you may find several species of *Begonia*.

Forest fruits also take astonishing forms and colours. *Tambourissa* fruit burst open to reveal orange flesh spangled with red and black seeds. "Ravenala" pods have black seeds surrounded by little cottony puffs of cobalt blue. It is time to speak of Madagascar's national symbol, the "traveller's palm" to foreigners, *Ravenala madagascariensis* to botanists, and word for word to Malagasy "the leaf of the forest".

The tree's leaves resemble banana leaves, but unlike bananas it has a trunk. The leaves grow in a single plane, opened like a gigantic fan, and they accumulate water in their axils. The thirsty traveller is supposed to tap them for fresh water, which gives the tree its European name. In fact the water squirms with mosquito larvae, insects and other forms of life, so it is fairly unappetizing in these regions where clean water rushes in streams down every hillside. The "ravenala" is planted in tropical gardens for its majestic silhouette, and thus joins its Amazonian cousin *Ravenala guyanensis*, (or better *Phenakospermum guyanensis*), which has dark red arils, but is less stately in shape. Both Malagasy and Amazonian plants are sun loving trees. The Malagasy form grows in clearings within the rainforest, and also in immense colonies on the cleared foothills of the East up to 600 m altitude. The trunk and leaves are the main materials of eastern houses; the seed can be ground to edible flour. The intense blue of the aril remains unexplained.

With the "ravenala" we enter the secondary forest, called the "savoka", which grows after the primary forest is destroyed. In 1960, Guichon estimated 3,584,000 ha of "savoka", for 6,396,000 ha of primary forest. We can only be pessimistic about their future.

After the felling of the forest, and then the abandonment of the tavy clearing, a thicket springs up. It is composed of huge herbaceous plants, with ginger and ferns, of vines: *Rubus*, *Flagellaria*, *Mikania*, and of bushes. This gives way to a later stage dominated by *Ravenala, Haronga*, bamboos, etc.

Many of these are foreign species: *Rubus moluccanus* comes from Asia, and forms great draperies giving huge insipid "raspberries". *Psidum guajava*, comes from South America, as does *P. cattleianum*, in spite of its common name "Chinese guava". *Lantana camara*, *Solanum auriculatum*, *Clidemia hirta* have become well-established constituents of the savoka.

The success of these foreign species probably results from the fragility of island flora and the absence of Malagasy plants specialized for niches in disturbed habitats. After all, man only arrived recently, and before that the forest was essentially "virgin". Sun loving plants only grew in the rare clearings.

On the coast, the "savoka" is full of *Philippia* and *Psidium*, which shows that the soil is extremely degraded. It seems clear that a single cycle of "tavy" cultivation is no handicap to eventual forest regrowth. In contrast, repeated clearing at short intervals does not allow soil regeneration, and leads to irreversible degradation. You end with the open landscape usual in the coastal zone, with banal grasses like *Imperata cylindrica*, *Axonopus compressus*, etc. and huge colonies of ferns, particularly *Pteridium aquilinum*.

The distribution of Malagasy humid forest is surprising. It seems normal to find it on the wet east coast, the island's Windward Domain. It is odd to find large extension of humid forest to the northwest, in the Sambirano Basin, between the savannahs of the far north and those of the west; the mountain massif of the Tsaratanana explains this climatic anomaly. The Sambirano forest is quite distinct, with marked endemism. Its plant formations are more heterogeneous than those of the east. We may consider these forests more as a transition zone rather than a barrier between the two savannahs, with interdigitation of forest and grassy milieux which facilitates species transfer.

This zone is among the richest in plant life, in spite of its monotonous appearance to superficial observers. In particular it has many edaphic variations, including marshes and swamp forests peopled with *Pandanus* and *Raphia*, ferns and Cyperaceae, and the remarkable Aracea *Typhonodorum lindleyanum*.

Raffia deserves special mention: its products are exported throughout the world. The gigantic palm leaves may reach 10 – 12 m long, growing upright in an opulent plume. Unfortunately, the species is overexploited. Besides the fibres it gives wax, alcohol, building wood . . . and the terminal bud can be eaten. Raffia reproduces only slowly, and flowers just once. The terminal bud gives off two or three enormous

Fig. 2.14 *Nepenthes madagascariensis*, the carnivorous pitcher plant of the southeastern marshes (A. Jolly)

Fig. 2.15 Marsh on the east coast, with *Typhonodorum lindleyanum*, palms and pandanus (Ph. Oberlé)

pendant inflorescences, giving thousands of hard-covered gleaming fruit, then the tree dies. Unthinking raffia cutting can suppress flowering and stop reproduction.

At the other end of the island, near Fort Dauphin, grows another strange marsh plant. The midrib of the *Nepenthes madagascariensis* leaf is prolonged into a vase, surmounted by a small cover, and containing a colourless liquid. When the cover lifts the vase fills with rainwater. Various bacteria and insects enter and die there (or sometimes live there). This "trap" has earned *Nepenthes* the reputation of being carnivorous, though it has not yet been shown scientifically what use the plant makes of its prey.

There is a popular belief that tipping out the water from the bases will bring rain — a belief reported by Flacourt as early as the seventeenth century. *Nepenthes* is a fine example of Asiatic distribution with 75 circum-Pacific species, of which two live in Madagascar and one in the Seychelles. We might point out an odd error on a Malagasy stamp which shows *N. madagascariensis* under the name of *N. pervilleana*, the Seychelles form.

The lakes, lagoons and canals of the East contain a pantropical flora which is interesting, but not unique to Madagascar. We can deplore the invasion of the water hyacinth (*Eichornia natans*), which comes from Amazonia. It is actually quite beautiful when it flowers, but this plant, introduced for ornamental ponds, has proliferated all over Africa and Madagascar covering rivers and canals so thickly as to stop navigation. Its local name "tetezanalika" means "dog's bridge", and does describe its density. People have calculated that ten "mother plants" can turn into a million in less than a year! The water hyacinth is all too clear an example of the rupture of an evolved equilibrium by an introduced foreign species.

Fig. 2.16 Grove of Pandanus trees and cattle pen near Tamatave (W. Ellis, 1858)

THE COASTAL FLORA

The flora of both east and west coasts contains many widely-distributed species: *Ipomocea pes-caprae* and *Canavalia obtusifolia* which are pantropical, *Plumbago aphylla* and *Pedalium murex*, which are Pacific, and *Thuarea involuta* which is paleotropical. There is some antagonism between two Goodenlaceae, *Scaevola plumieri's* Atlantic distribution reaches to the west coast of Madagascar. *S. serica*, which comes from the Pacific, is common on the east and rare on the west coast, and has only recently spread to African shores.

The circum Pacific distribution which we have already seen in *Nepenthes* reappears in the forest which grows on the east coast sands. The genus *Cycas*, with the probably endemic species *C. thouarsii*, does not extend beyond these Windward coast forests, like *Barringtonia butonica*, *Tournefortia argentia*, and others.

Cycas thouarsii is a Prespermaphyte, a plant so primitive it does not produce seeds. It looks much like a palm, and seems to be the ''anthropophagous tree'', described by early travellers. According to their tales, the Malagasy sometimes offered a young virgin to this monstrous plant, and illustrations show the unhappy victim writhing in the convulsive clutch of its leaves! We know nothing of the origins of these inept beliefs. The mangrove community will be described in detail in the next chapter.

Let us then leave the world of the Malagasy flora, with this final view of the coasts. It is a unique world, with the characteristics of an island and also of a continent, separated from Africa perhaps a hundred million years ago, having formed a part of the vast southern land mass of Gondwana. From the earliest Gondwanaland elements, and later invaders, rich and unique flora has evolved over millenia. The characteristics of the island: its climatic differences, its watersheds, its broken topography, and its diversity of soils have favoured the differentiation of plant taxa.

Originality, beauty, and scientific interest — everything joins to make us respect, study and love the vegetation of Madagascar, at the level of institutions as well as individuals. These few pages can only be a modest introduction — and perhaps an incitement to deeper knowledge of the plant world of Madagascar.

Fig. 2.17 *Ravenala madagascariensis*, the Travellers' Palm, symbol of Madagascar (A. Jolly)

REFERENCES

Flore de Madagascar, collection publiee par le Museum national d'Histoire naturelle, fondee en 1936 par H. Humbert, dirigee par le professeur Jean-Francois Leroy. A ce jour, 136 familles de plantes de Madagascar ont ete publiees.
Abadie, 1951 – 52 Inventaire des especes fruitieres comestibles a Madagascar, *Bull. de l'Acad. Malg.*, XXX, 185 – 204.
Baker, J.G. 1886 *Flora of Madagascar*, London.
Baillon, H. & Drake, E. del Castillo Histoire naturelle des plantes dans la Collection de l'*Histoire physique, naturelle et politique de Madagascar*, publiee par A et G. Grandidier.
Baron, Rev. 1901 – 06 Compendium des plantes malgaches. *Revue de Madagascar.*
Boiteau, P. 1937 Richesses floristiques et fauniques de Madagascar. *Revue de Madagascar*, n° 20, oct., 53 – 74.
Catala, R. 1936 *Dans le Sud de la Grande Ille, Revue de madagascar*, n° 14, avr., 79 – 132.
Decary, R. 1930 La Flore de la ville de Tananarive, *Bull. Acad. Malg.*, t.XIII, 127 – 49.
Fournols, J. 1935 La Sylve malgache, *Revue de madagascar*, n° 11, juill., 109 – 44.
Francois, E. 1937 Plantes de madagascar, *Memoires de l'Acad. Malg.*, CCIV, 71pp. 23 pl.
Francois, E. 1941 Les mille aspects de la flore de Madagascar, *Revue de Madagascar*, n° 30, juill., 7 – 45.
Gueneau, P. Bois et essences malgaches, *Bull. de Madagascar*, n° 304 (Sept. 1971), 305 – 06 (Oct./Nov. 1971), 311 (Avr. 1972).
Humbert, H. 1927 La destruction d'une flore insulaire par le feu, principauz aspects de la vegetation a Madagascar, *Memoires de l'Acad. Malg.*, V, 79 pp. 41 pl.
Humbert, H., Leandri, J. 1954 Cinquante ans de recherches botaniques a Madagascar, *Bull. Acad. Malg.*, special cinquantenaire, 33 – 42.
Koechlin, J., Guillaumet, J.-L., Morat, Ph. 1974 *Flore et vegetation de madagascar*, Vaduz, Cramer ed., 687 pp. 187 ill.
Louvel, 1931 *Les plantes ornementales et curieuses de madagascar.*
Millot, 1964 L'ethnobotanique malgache, *Civilisation Malgache* 1, Fac des lettres et sciences humaines, Tananarive, 15 – 23.
Paulian de Felice L. 1958 Les Orchidees malgaches, *Revue de Madagascar*, 2° trim., n° 2, 47 – 52.
Perrier de la Bathie, 1921 La vegetation malgache, *Annales du Musee colonial de Marseille*, 268 pp.
Perrier de la Bathie, 1927 *Le Tsaratanana l'Ankaratra et l'Andringitra. Memoires de l'Academie Malgache*, III, 68 pp.
Perrier de la Bathie, 1936 Biogeographie des plantes de Madagascar, *Soc. d'Ed. Geogr. Col.*, Paris, 156 pp. 40 pl.
Poisson, H. 1948 Les catalogues et nomenclatures de plantes malgaches. *Mem. de l'Acad. Malg.*, fasc. hors serie, 157 – 61.
Reynolds, G.W. 1958 Les Aloes de Madagascar, *Suppl. au Naturaliste Malg.*, C, 156 pp. 103 fig., 18 ph. col.

CHAPTER 3

Flora of the Malagasy Southwest

RACHEL RABESANDRATANA

Suppose a tourist, a botanist or a naturalist wants an impression of the flora of Madagascar. He may visit a large part of the island but make excuses to himself that he simply hasn't time to visit the Malagasy Southwest. Believe me, he has seen nothing.

On the coast you meet an abrupt contrast between evergreen mangroves growing by the sea's edge and the grey tinged "bush" or thornscrub immediately adjacent on solid ground. The bizarre plants of the bush are adapted not only to severe climatic conditions, but also to very varied soils. It is not surprising, therefore, that the Malagasy southwest contains not only species, but genera and even families which are wholly endemic, such as the Rhopalocarpaceae, and the extraordinary Didiereaceae. Perrier de la Bathie estimated that 48% of the genera are unique to this single region! And besides its purely scientific interest, the bush constitutes a reservoir of medicinal plants in everyday use.

I. THE CLIMATE

The flora of the southwest belongs to what Perrier de la Bathie called the Leeward Zone. Humbert placed it in his Southern Domain, which stretches along the coast from Morombe to Fort Dauphin.

The southwest lies at low altitude, about 0 – 400 m. The climate on the coast is what Morat terms subarid, while in the interior it is semi-arid. The Tropic of Capricorn divides the region just south of Tulear.

Annual rainfall averages some 350 mm. However, there are years of real drought. In 1970, Tulear received only 158 mm of rain which fell during a mere 15 days. In the twenty years from 1960 to 1980 the highest rainfall occurred in 1968 with 724 mm spread out over 53 days.

There is no clear-cut wet season, according to Riquier's definition. In general the months of December, January and February receive more than 50 mm of rain, while the months of April to October have less than 40 mm. Precipitation is very irregular.

Cold currents parallel the coast and force moisture out of humid air, which then condenses over the sea. This increases the aridity of the winds reaching the land. Even so, atmospheric humidity plays a large role in the southwest. In the "dry season", that is, outside the months of December, January and February which we consider "rainy", nightly dew may be a considerable source of water for plants.

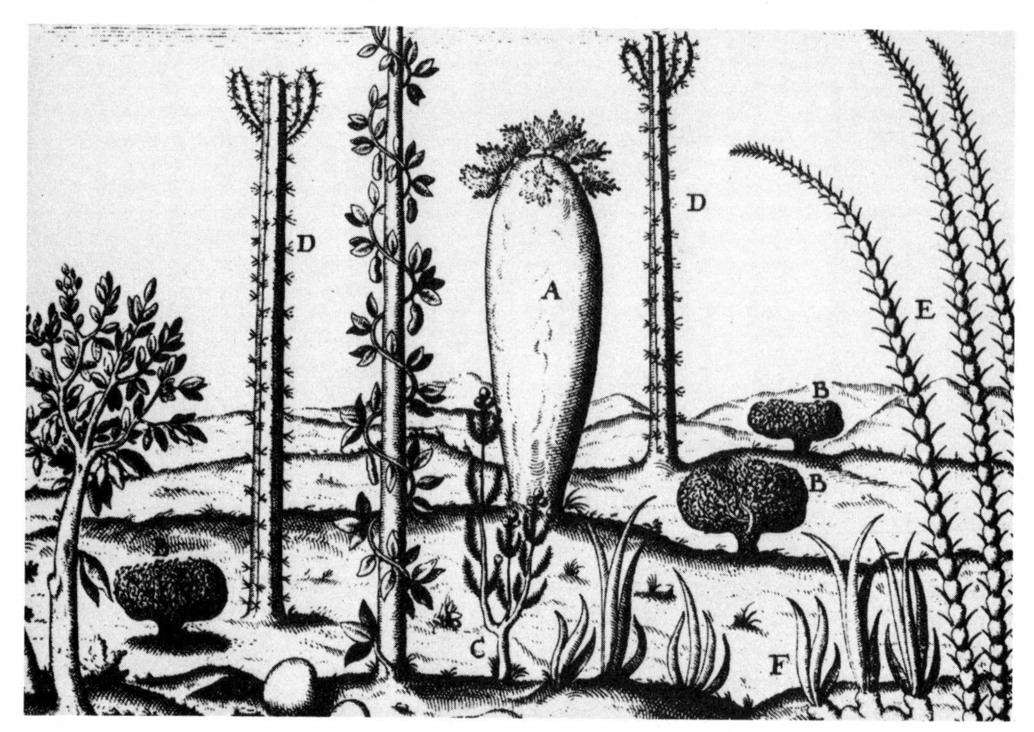

Fig. 3.1 Plants of the Southwest (De Bry, 1601)

Fig. 3.2 Contrast between the evergreen mangrove forest and the greyish "Bush" immediately beside it on dry land (H. Rabesandratana)

II. SOILS

Various types of soil make up the southwest:
— calcimorphic soils which form a large part of the calcareous Mahafaly Plateau;
— red sands, ferruginous soils which originate from sandstone;
— primitive soils like the coastal sand dunes;
— alluvial soils, with a shallow water table;
— saline soils including soils on ancient salt crusts, swampy soils, beach and mangrove soils.
There are also intermediates between these basic soil types.

III. VEGETATION

Vegetation is the reflection of soil and climate. It is difficult to draw a precise line between the dense dry woodlands of the West and the southwestern woodlands because one grades progressively into the other. However we can consider the true bush of the southwest as a limiting type, which reveals the pressure of arid conditions.

In the bush there are small bushes 1 – 2 m high and trees, usually 3 – 4 m high, occasionally as tall as 10 m. In the difficult conditions of survival, plants fight the lack of water in various ways. Many species have deciduous leaves, much reduced leaves, or no leaves at all. This results in lack of shade, and thus little stratification of plant layers. If from time to time an arborescent stratum dominates the whole, it is still sparse. Even so, Koechlin defined several types of bush as a function of height:
— high bush with trees which can reach 8 – 10 m in height, the upper stratum largely composed of Didiereaceae emerging above an extremely dense bushy layer;
— high bushy scrub, formed of bushes of various sizes, interlaced with each other and reaching as much as 5 m in height. Various Didiereaceae and leafless Euphorbs grow in this formation;
— low bushy scrub, where the layer of bushes reaches only 1 – 2 m in height.

Within each of these structural types of plant community there are different formations where different species, genera, or even families predominate.

As early as 1912, Poisson laid particular emphasis on the great plant families found in the southwest, in his "Recherches sur la Flore Meridionale de Madagascar". Basse (1934) as though to underline the importance of the southern families and genera called the various formations "Didiereaceae bush, leafless Euphorb tree bush, Pachypodium bush, Adansona bush". Her classification is useful for these plants impose themselves on the view both by their number and their large size.

Let us now make the acquaintance of these major families.

A. *The principal families*

1) Didiereaceae

This Malagasy family is endemic to the southern Domain. It includes 4 genera: *Didierea*, *Alluaudia*, *Alluaudiopsis* and *Decarya*.

The Didiereaceae are woody plants, most of them succulents, highly adapted to withstand aridity. They grow long branches bearing deciduous leaves that fall after only a brief life, and short branches bearing thorns, leaves, and flowers. Keraudren thinking of the Didiereacea called them "serpent trees".

a) Genus *Didierea* or Sony. The fleshy branches have large aquiferous cores.

Didierea madagascariensis Baill. or "octopus tree". This tree has a short trunk and long upright branches. It reaches 4 – 10 m in height. A few shorter twigs grow into vertical ramifications at the top, while the long branches are fat thorny stems, bearing elongate leaves in rosettes of 6 to 10. The short twigs appear at the axils of these leaves. In turn these twigs carry spines of unequal length, in cross pattern, and a rosette of 4 to 5 leaves which differs from those of the long branch. In male individuals the twigs are elongate, reaching 6 – 7 cm in length. In females, these twigs are reduced to mere cushions on the branch. This species grows on decalcified red sands along the coast between Morondava and Lake Tsimanampetsotsa.

Didierae trollii Capuron and Rauh., or sonibarika. This has an extraordinary pattern of growth. In the young tree a snarl of horizontal branches crawl over the surface of the soil. Upright trunks emerge from these which curve over in turn to form successive levels of further horizontal growth. The twigs

Fig. 3.3 *Didierea madagascariensis* or octopus tree (H. Rabesandratana)

Fig. 3.4 *Didierea trollii* (A. Jolly)

bear a smaller number of spines and leaves than *D. madagascariensis*. There are no differences between male and female individuals. Its range extends from Lake Tsimanampetsotsa to the Mandrare River.

b) Genus *Alluaudia*, which has many varied forms.

Alluaudia ascendens Drake, or songo be. This is the largest of all the Didiereacea, a tree which reaches 15 m in height. Its trunk is thick but short, culminating in a group of nearly vertical branches. Its reddish bark carries blunt spines like the knobs on a mace, which circle the trunk along with the leaves, in very vertical helices. One can recognize this species at a distance by the inflorescences like muffs along the upper portion of the branches. *A. ascendens* is confined to the basin of the Mandrare River.

Alluaudia procera Drake, the fantsiholitra or, again, octopus tree. In its early years this plant grows like a bush. Later its mature tree shape appears, with thorny branches raised to the sky, slightly curved, recalling a half-opened fan. The branches in fact resemble the arms of a squid rather than an octopus, quasi parallel. The bark is greyish, encircled by helical rows of sharp, close-set thorns. It can be identified at a distance by the spherical terminal inflorescences balancing on the top of each branch. This is the only truly woody Didiereacea. Its wood is light, resistant, and does not rot, so it is used in the south for house building. It seems to grow in regions of crystalline bedrock more or less covered with sandy soil.

Alluaudia montagnaii, Rauh. possesses a trunk which is thick at the base, but divides quickly into 3 or 4 nearly vertical branches that narrow progressively toward their tips. The ends, which become pliable or fragile, bend over under the effect of wind. This results in an easily recognizeable shape. The species grows in the neighbourhood of Itampolo.

Alluaudia humbertii Choux. This is a bushy species, whose branches grow both horizontally and vertically. In the Ihosy region it grows on soils over crystalline bedrock.

Alluaudia dumosa Drake or rohondra. This is a dark, bushy tree, with thick cylindrical branches and characteristic candelabra shape. It lives on silicious soils in the south from Ampanihi to Fort Dauphin, including the region of Beraketa.

Alluaudia comosa Drake or sonoratsy. This tree has a short trunk, and the branches stop at the top in a flat plane. Thus, the whole tree is shaped like a funnel. This characteristic silhouette is easy to recognize even at a distance. This species is an indicator of calcareous soils, from Tulear to Fort Dauphin.

Fig. 3.5 *Alluaudia comosa* funnels out of the ground like a vegetable tornado (H. Rabesandratana)

c) Genus *Alluaudiopsis*. This differs from *Alluaudia* and *Didierea* by growing as bushes, not trees. *Alluaudiopsis marnierian* Rauh with its carmine red flowers grows together with *D. madagascariensis* on decalcified red sands. *Alluaudiopsis fiherenensis* Humb. and Choux with its yellow flowers grows on calcareous soils like *A. comosa*.

d) Genus *Decarya*. There is just one species, *Decarya madagascariensis* Choux. This is a bushy tree, whose thorny branches grow in right-angled zig-zags. It grows in the region between Ampanihi and Ambovombe.

Fig. 3.6 *Euphorbia oncoclada*'s branches are like linked sausages (H. Rabesandratana)

Fig. 3.7 *E. oncoclada* in flower (H. Rabesandratana)

2. *Euphorbiaceae*

The genus *Euphorbia* or famata can hardly be avoided in the Malagasy southwest. ''Coralliform'' species recall the branching madrepores of the nearby barrier reef.

Euphorbia laro Drake, or laro, with its thin, cylindrical branching twigs grows on alluvial soils. Its latex can ulcer the skin or even blind.

Euphorbia stenoclada Baill. or famata botribotrika, endemic to Madagascar, is a bush or shrub with scaly trunk and secondary twigs that terminate in spines. This leafless species can be recognized by its greyish-white colour and its generally flat-topped or sometimes spherical shape. Resin from the trunk serves to caulk pirogues. Zebus eat the extremities of the branches. This species grows on calcareous Quaternary sands.

Euphorbia oncoclada Drake or famata betonato, is endemic to Madagascar. It is a grey-green bush without thorns, no more than 30 m tall. The twigs have constrictions which correspond to branching points with their terminal buds. Each articulation is spotted with brown dots which correspond to the leaf-scars left by deciduous leaves. The whole thing looks like linked strings of sausages. Seen from a distance the bush is roughly spherical. This species grows on calcareous soils.

Euphorbia leucodendron Drake or befotsy, endemic to Madagascar. This euphorb is yellow-green, thornless, with branches like sausages linked end-wise in groups of three. It grows on calcimorphic soils.

Euphorbia fiherensis H. Poisson, endemic to Madagascar. This is a non-thorny shrub, with thin green branches. It is parasol shaped, and grows at the foot of calcareous cliffs in the Tulear region.

Euphorbia plagiantha Drake has neither leaves nor thorns. It is yellow-green with extremely elongated, pointed, fleshy branches. The trunk is scaly. This shrub lives on sands at the foot of calcareous cliffs in the Tulear region.

Euphorbia antso H.Bn. This bush bears deciduous yellow-green leaves grouped at the ends of the branches. Like the preceding species it grows on calcareous sands at the foot of cliffs in the Tulear region.

Euphorbia enterophora Drake or betinay, endemic to Madagascar. This is the largest of the euphorbs, and may reach 5 m in height in the dry forests of the west. It is easily recognized by its parasol shape and by its branches which are flattened in cladodes.

3. *Genus Pachypodium (Apocynaceae).*

There are four species of pachypodium, the "elephants foot", in the south. They are all single-stemmed trees unlike the cushion-shaped pachypodium species which grow on rocks elsewhere in Madagascar.

Pachypodium geayi C. & B. or vontaka. Its spines grow in groups of three on the trunk. This species can be recognized by its upright shape, its numerous long leaves which form bouquets on the end of short branches, and its white flowers. The double fruit is composed of two opposed follicles, each easily 30 cm long. It grows on calcareous soils between Tulear and Ambovombe.

Pachypodium lamerei Drake or hazo tavoahangy. This species differs from the last by its swollen base. Its spines grow in groups of three.

Pachypodium rutenbergianum Vatke. This tree has an upright trunk, with spines in groups of two. The species grows mainly in the dry deciduous forests of the northwest, but may also be found south of Tulear.

Pachypodium meridionale (H. Perr.) M. Pichon is a tree with spines in groups of two.

4. *Genus Adansonia (Bombacaceae).*

The seven Malagasy species of baobab live in the Leeward Domain. Three of them grow in the south where they are dominant elements of the southern Bush.

Adansonia fony J. Bn. or little boabab, is a tree 2 – 5 m tall, with monstrous trunk and with leaves divided into fine-toothed leaflets. The flowers are brilliantly coloured: orange petals and numerous yellow stamens. It usually grows on red sands, but occasionally on calcareous sand. North of Tulear there is a lovely forest of *Didierea madagascariensis* and *Adansonia fony* on red sand.

Adansonia za H. Bn. This species reaches 15 – 20 m height. The pointed leaflets have petioles; the flowers are yellow. It grows only in the Mandrare River basin, in association with *Alluaudia ascendens.*

Adansonia madagascariensis H. Bn. is a rare species, 20 – 25 m tall. The leaves have petioles but the leaflets are sessile and spatulate.

5. *The Leguminaceae.*

The members of this family can hardly be missed. The most remarkable is *Baudouinia rouxevellei* H. Perr. (Cesalpinaceae). Its Malagasy name, manjakabentañy, means "great king of the earth". This tree is localized between the valleys of the Fiherenana and the Onilahy rivers, on calcareous bedrock. The wood of the trunk is deeply canalized and very hard. Once stripped of the bark it is used to make traditional canes and modern lampstands, sold to tourists as "bois sacre".

Delonix adansonioides (R. Vig.) R. cap. has a swollen trunk, resembling a baobab, but highly constricted at ground level. It can grow by budding. It has an odd attribute: branches planted side by side thicken as they grow and can fuse to form a continuous single-tree enclosure. Such enclosures can be seen at Itampolo and at Efoetsy, not far from Lake Tsimanampetsotsa.

Euphorbia laro Drake, or Laro, with its thin, cylindrical branching twigs grows on alluvial soils. Its latex can ulcer the skin or even blind.

Euphorbia stenoclada Baill. or famata botribotrika, endemic to Madagascar, is a bush or shrub with scaly trunk and secondary twigs that terminate in spines. This leafless species can be recognized by its greyish-white colour and its generally flat-topped or sometimes spherical shape. Resin from the trunk serves to caulk pirogues. Zebus eat the extremities of the branches. This species grows on calcareous Quaternary sands.

Euphorbia oncoclada Drake or famata betonato, is endemic to Madagascar. It is a grey-green bush without thorns. Various *Cassia*, *Bauhinia*, *Albizia*, *Acacia*, *Mimosa* and *Dicrostachys*, to cite only the commonest genera, fill out the list of large leguminous trees of the southwest.

Fig. 3.8 *Pachypodium geayi* has a tall straight trunk with follicles (H. Rabesandratana)

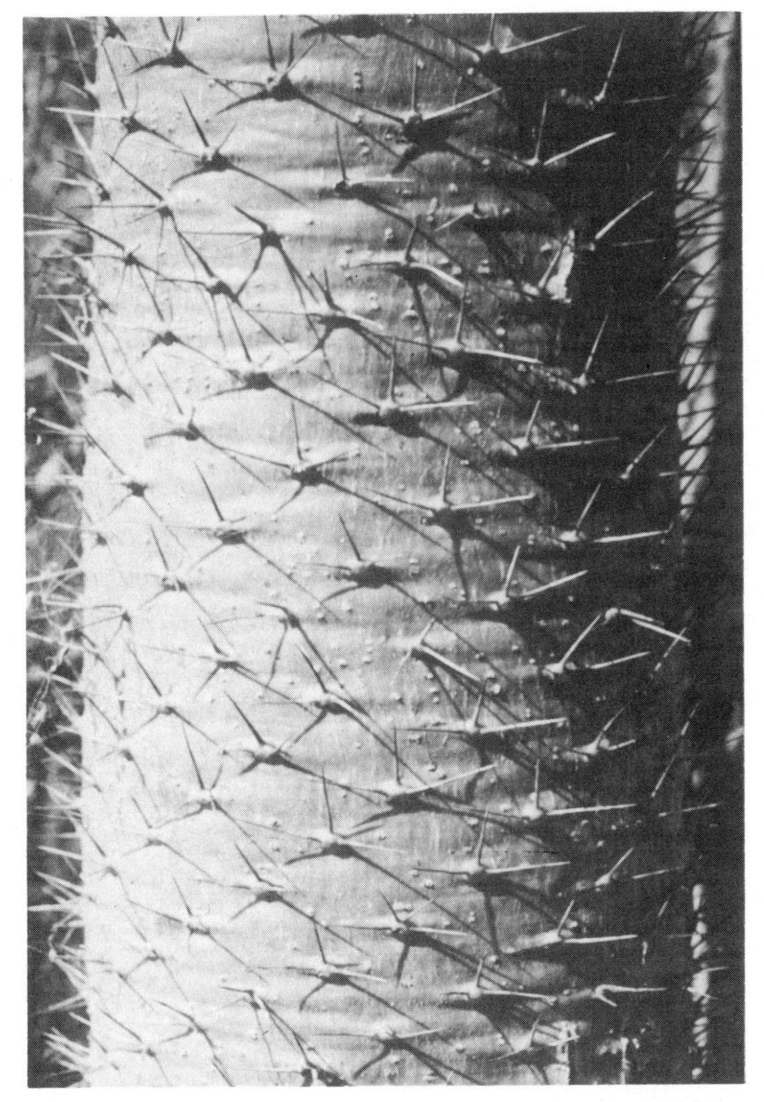

Fig. 3.9 *P. geayi*'s spines are grouped in triplets on the trunk (H. Rabesandratana)

B. CONVERGENT ADAPTATIONS

The struggle against water loss governs all plant growth in the south. Thus the vegetative parts of plants show many convergent adaptations, which result in similarities of appearance.

1. *Root adaptations*

Besides geophytes with true bulbs, like *Gloriosa virescens*, there are many other annuals and perennials with swollen roots. In one lianescent Papilionacea, *Dolichos fangitsy*, the root which stores water may be as large as a coconut. On the Mahafaly Plateau thirsty goatherds seek them out. The goatherds tap the ground with a stick, and find the tubercules by their characteristic resonance.

Fig. 3.10 Various bottle trees planted as sacred vegetation by a Mahafaly tomb (Ph. Oberlé)

There is also great disproportion between the enlarged root system and puny vegetation above-ground among various Asclepiadaceae, Cucurbitaceae, Dioscoreaceae, Euphorbiaceae and Vitaceae.

2. *Trunk adaptations*

Pachycauly reigns in this arid region of the Island, to the point where species of very diverse families have swollen trunks, like barrels which store the plants' indispensable water. Laymen call all these different species "bottle-trees".

a) Anacardiaceae: *Operculicarya decaryi* has a thick trunk, though often of small size.

b) Apocynaceae: All four *Pachypodium* species described above have thick, fleshy, thorny trunks, gorged with water.

c) Bombaceae: The three *Adansonia* species have huge spongy trunks filled with water which can slake the thirst of men and animals. Men drink by "tapping" a hole into the bark. Sometimes the entire tree is cut down and split in two to form a watering trough for zebu.

d) Burseraceae: *Commiphora monstruosa* has much reduced leaves and a swollen trunk, fixed solidly on the soil by roots which spread out on the surface.

e) Didiereaceae: Trunk and branches of most Didiereaceae are large and filled with spongy pulp.

f) Euphorbiaceae. *Euphorbia antso* has a thick, subcylindrical trunk, ringed with circular scars. *Givotia madagascariensis* has a very thick trunk of water-resistant wood which is used to make pirogues.

g) Hernandiaceae. *Gyrocarpus americanus* has palmate leaves, resembling those of *Jatropha mahafaliensis*, as well as those of *G. madagascariensis* all species which grow in the Tulear region.

h) Leguminoseae: *Delonix adansonioides* has a swollen baobab-like trunk which gives its scientific name.

i) Moringaceae: *Moringa drouhardii* also recalls a baobab, but the compound tripennate leaves possess petioled leaflets.

j) Passifloraceae: The rare *Adenia* species are lianescent forms which grow at the base from a short, bulbous trunk.

k) Vitaceae: Another liana, *Cyphostemma laza* has a very bizarre appearance. It has a conical trunk some 2 – 3 m tall, from the top of which grows a whip-like stem that bears the leaves. This plant grows in the south but is commoner in the Western dry woodlands.

Water can also be stored, in less striking fashion, in fleshy twigs and branches. They offer a multitude of recipients of more modest size than a single swollen trunk. For instance, *Notonia descoingsii* (Compositae), the coralliform *Euphorbia* species already listed, *Alluaudia tumosa* (Didieraceae), lianescent Asclepiadaceae such as *Cynanchum* species, the wild Malagasy vanillas, and *Folotsia grandiflorum* all have fleshy branches.

Another means of combating water loss, besides storing water, is to reduce length of branches or to turn them into spines. This occurs in *Rhigozum madagascariensis* (Biononiaceae), *Commiphora simplicifolia* (Burseraceae) which can be recognized by its horizontal branches with sympodial branching, and *Megistostegium nodulosum* (Malvaceae), easily recognized by its large pink flowers called "Cape roses".

3. *Leaf adaptations*

a) Thickened succulent leaves also store water. This common phenomenon occurs in:

Compositae: *Senecio* species.

Crassulaceae: *Kalanchoe beharensis* with leaves shaped like elephants' ears, and *Kalanchoe grandidieri*.

Cucurbitaceae: *Xerosicyos danguyi*, a lianescent plant with coin-shaped leaves which grows on calcareous sands and *X. perrieri* much like the former but with smaller leaves.

Liliaceae: There are many endemic Aloes with upright stem and large thick juicy leaves that grow in a terminal rosette. Among them we may note *Aloe vaotsanda*, *A. divaricata*, 1 – 2 m high and with flowers in multi-stemmed bunches, *A. vaombe*, sometimes 5 m high with non-ramified stem and downcurved leaves of a fine dark green below the rose-madder flowers, which covers vast areas of the Antanimora region, *A. helenae* with single stem 3 m tall and scarlet and yellow flower spikes, and finally the extremely rare *A. susannae*, with stem 5 m tall and an inflorescence of lemon yellow flowers which may itself be 2 m high. This last species grows only in the Mandrare basin, and is probably near extinction.

Solanaceae: *Lycium tenuum* has small fleshy leaves.

Zygophyllaceae: *Zygophyllum depauperatum*.

b) Microphyllia, or reduction of leaf size, is another common way to prevent water loss. We may cite among many:

Operculicarya decaryi (Anacardiaceae) with miniscule composite leaves, *Rhigozum madagascariensis* (Bignoniaceae) which is easy to recognize by the galls which frequently deform its branches, *Albizia atakataka*, *Bauhinia grandidieri*, *Mimosa delicatula* (all Leguminosae), *Holmskoldia microphylla* (Verbenaceae), and *Terminalia ulexoides* (Combretaceae).

c) In some cases leaves and stipules are transformed into thorns. The frequent prickly plants do not exactly ease penetration into the plant communities of the south. One rapidly overgeneralizes and declares that all the plants are thorny, and this region is sometimes called the spiny desert. However, our samples indicate that a mere 10% of plant species are thorny, though these cover large areas. Among thorny plants let us note *Barleria lupulina*, with yellow flowers, *B. her* which forms a thorny bush with red flowers, *Blepharis calcitrapa* with violet flowers (all Acanthaceae), the *Pachypodium* species (Apocynaceae) and Didiereaceae already listed, compositae like *Dicoma grandidieri* with needle-shaped leaves,

Fig. 3.11 *Xerosicyos danguyi* has coin-shaped leaves (H. Rabesandratana)

Leguminosae like *Acacia* species, and Salvadoraceae like *Azima tetracantha*, which, as usual, is endemic to Madagascar.

d) Leaves are frequently deciduous in the dry season. This is the case in *Poupartia caffra* (Anacardiaceae) in the southern savanna, of *Didierea madagascariensis*, of Euphorbs like *Jatropha mahafaliensis*, and *Givotia madagascariensis*, Hernandiaceae like *Gyrocarpus americanus* and Pedaliaceae like *Uncarina stellulifera* the harpoon burr. Botanical identification and collecting is admittedly difficult in the dry season for the naked trees and bushes all resemble each other. Only experience allows one to distinguish, for example, the small trees *U. stellulifera*, *J. mahafaliensis* and *G. americanus* when they have lost their leaves.

e) These last three species in the hot season hang their leaves down limply, from 9 o'clock in the morning. The leaves do not rise again to their normal horizontal position until dusk. This phenomenon of drooping leaves is certainly caused by water deficit, but also serves to protect the leaves from direct rays of the sun. The same thing occurs in *Kosteletzkya diplocrater* (Malvaceae), various *Grewia* species (Tiliaceae), and *Croton* species (Euphorbiaceae).

f) A waxy coating on leaves and stems reduces water loss. You can often find hard empty sheaths of wax on the ground which encircled *Kalanchoe* stems. All above-ground parts of Euphorbes and Alluaudia are also waxy. *Psiadia altissima* stick to the drying paper with their wax when you attempt to prepare herbarium specimens.

Hairy coatings also decrease evaporation from the vegetative parts of several *Kalanchoae*, Malvaceae like *Megistostegium modulosum*, Sterculiaceae like the *Dombeya* species, and Amaranthaceae like *Aerva javanica* and *Aerva tiangularifolia*.

IV. FLORAL COMMUNITIES

The floral communities of the southwest can be classified by the soils on which they grow.

1) *Vegetation of the calcareous plateau*: Bush with *Alluaudia comosa*.

We have gathered an average of 70 plants at each sampling site in the *A. comosa* bush of the Tulear region. A certain number of species appear in almost every sample.

Terminalia divaricata, *Asparagus vaginellatus*, *Jatropha mahafaliensis*, *Tallinella grevei*, *Cynanchum* sp., *Xerosicyos danguyi*, *Polycline proteiformis*, *Euphorbia oncoelada*, *Rigozum madagascariensis*, *Blepharis calcitrapa*, *Secamone* sp., *Commiphora monstruosa*, *Croton* sp., *Commiphora* sp., *Holmskoldia microphylla*, *Alluaudiopsis fiherenensis*, *Zygophyllum madecassum*, *Vernonia* sp., *Kalanchoae* sp.

2) *Undeveloped soils formed of white sand: Bush with Euphorbia stenoclada, H. Baill.*

These white sands constitute a cordon of dunes more or less along the sea. The dune vegetation is very open bush grading down to beaches with no woody vegetation. Plants cover only about 25% of the soil surface. Average height of the bush is only 0.5 – 1.5 m, though a few sparsely distributed plants like *Euphorbia stenoclada* reach 4 – 6 m. This community corresponds to Koechlins "low bushy scrub".

Florisitic inventory of four sample sites shows that of 112 species collected, only 24 are present in as many as three sites. The other species are not just there at random, but indicate the stage of soil evolution; some belong properly on decalcified red sand, some on the calcareous plateau.

A certain number of plants are always present; *Euphorbia stenoclada*, *Salvadora angustifolia*, *Maerua filiformis*, *Mimosa* sp., *Poivrea grandidieri*, *Ecbolium linneanum*, *Helinus ovatus*, *Secamone cristata*, *Paederia* sp., *Setaria humbertiana*, *Lasiosiphon decaryi* var. *littoralis*, *Megistostegium nodulosum*, *Commiphora simplicifolia*, *Bauhinia grandidieri*, *Mundulea pungeus*, *Zygophyllum depauperatum*, *Aerva triangularifolia*, *Mollugo decandra*, *Barleria her*, *Grewia* sp., *Asparagus* sp., *Cissus bosseri*, *Clerodendron globosum*.

3) *Vegetation on decalcified red sand: high bush with Didierea madagascariensis.*

This is Koechlin's "Dense dry Forest with *D. madagascariensis*".

Of 118 species we have collected 38 are constant. A few are ubiquitous: *Terminalia divaricata*, *Salvadora angustifolia*, *Grewia* sp., *Poivrea grandidieri*, *Cedrelopsis grevei*, *Bauhinia grandidieri*. Others have a tendency to live on sand: *Mimosa* sp., *Maerua filiformis*, *Ecbolium linneanum*, *Ipomaea* sp., *Commiphora simplicifolia*. Others prefer calcareous soil: *Jatropha mahafaliensis*, *Gyrocarpus americanus*.

There are some species found exclusively on the red sands: *Didierea madagascariensis*, *Ipomea* sp., *Mundulea* sp., *Barleria humbertii*, *Adansonia fony*, *Chadsia grevei*, *Folotsia grandiflorum*, *Grewia* sp., *Kigelianthe madagascariensis*, *Commiphora orbicularis*, *Terminalia gracilis*, *Viscum* sp., *Euphorbia* sp., *Stapelianthus montagnacii*, *Erythroxylon* sp., *Aloe divaricata*, *Rhopalocarpus similis*, *Pachypodium lamerei*, *Gibotia madagascariensis*, *Xanthoxylon decaryi*, *Diospyros latispatula*, *Kosteletzkya diplocater*, *Neohumbertiella decaryi*, *Boscia plantefolii*, *Strychnos* sp., *Vanilla* sp.

4) *Vegetation of alluvial soils. Phragmites mauritianus* and *Typha angustifolia* predominate.

Along the rivers a few woody species form gallery forests, including the banyan *Ficus cocculifera* ssp. *Sakalavarum*. Aerial roots sprout from the branches of the banyan fig, lengthen vertically downwards and thicken before reaching the ground thus forming crutches for the horizontal branches. The whole group together has a weird effect. The most famous, the banyan, at Miary 7 km north of Tulear, has become a tourist curiosity, for this single tree covers almost a hectare of ground. The root-crutches are so thick and so much interlaced it is like standing in a herd of cream-coloured giraffes, among whom one can no longer make out which was the original central trunk.

5) *Secondary vegetation.*

Most of the alluvial soils are cultivated. Weed species grow at the edges of the fields, including: *Aerva javanica* (Amaranthaceae), *Leptadenia madagascariensis* (Asclepiadaceae), *Tridax procumbens* (compositae), *Euphorbia hirta* and *Ricinus communis* (Euphorbiaceae), *Cynodon dactylon* (Graminaceae), *Leonotis nepetaefolia* (Labiaceae), *Argemone mexicana* (Papaveraceae), *Ziziphus mauritiana* (Rhamnaceae), *Datura stramonium* (Solanaceae), *Lantana camara* (Verbenacea).

As Perier de la Bathie originally pointed out, the secondary vegetation includes almost no species endemic to Madagascar while in undisturbed bush endemics form the vast majority.

6) *Vegetation of salty soils.*

The general grouping of halophile vegetation includes the flora of salty soils, salt marshes, beaches, and mangroves. Bigot has described a succession of key species on salty soils, as a function of decreasing salinity: *Salicornia pachystachya*, *Cressa cretica*, *Arthrocnemum indicum*, *Sporobolus virginicus* (L.), *Sclerodactylon macrostachyum*.

The community of *S. pachystachya* generally includes only this species for nothing else tolerates such extreme saline conditions.

C. cretica may be joined by bushes in sickly form like the composite *Pluchea grevei* and two Salvadoraceae *Azima tetracantha* and *Salvadora angustifolia*.

A. indicum often grows with *Sesuvium portulacastrum*.

The Graminaceous *S. virginicus* grows in meadows, sometimes extensive ones, on former mangrove soils which are still occasionally flooded with sea water.

The endemic graminaceous plant *S. macrostachyum* grows where there is a saline surface crust. It is found with *Salsola littoralis* (another endemic species), a creeping convolvulus *Ipomea pes-caprae*, *Spirobolus virginicus* and several bushes: *Pluchea grevei*, *Acacia* sp., *Hyphaena shatan* and *Cryptostegia madagascariensis*.

When the salt crust has not formed or when the altitude is slightly higher you find *Zygophyllum depauperatum*, *Salvadora angustifolia*, *Azima tetracantha*, *Lycium tenum* and *Atriplex perrieri* (an endemic Malagasy species).

7) *Vegetation of salt marshes.*

Includes: *Typha angustifolia*, the large fern *Acrostichum aureum*, a Combretacea which is a marginal mangrove *Lumnitzera racemosa*, and *Juncus maritimus*, *Paspalum vaginatum*, *Cyperus laevigatus*.

Fig. 3.12 *Rhizophora mucronata*: adventive roots like piles brace the mangrove in the mud

8) *Beach vegetation*.

Is almost entirely composed of cosmopolitan species. Nearest the water it begins with creeping plants: *Ipomea pes-caprae* (Convolvulaceae) with its goats-foot leaves and *Canavalia obtusifolia* (Papilionaceae) or beach bean. There are colonizers like *Pedalium murex* (Pedaliaceae), *Dactyloctenium capitatum* (Graminaceae), *Cyperus maritimus* (Cyperaceae) and an endemic Plumbaginacea, *Plumbago aphylla*. Above these, the vegetation thickens out with shrub species like *Lycium tenuum* (Solanaceae), and *Scaevola plumieri*. Finally you reach the first bushes: *Dodonia viscosa* which is not confined to the sea coast and *Thespesia populnea* (Malvaceae). Vines like *Marsdenia cordifolia* (Malvaceae) mix in with the bushes.

Fig. 3.13 *Bruguiera gymnorhiza*: elbow-shaped protuberences anchor this mangrove

9) *Mangrove communities.*

The Greek Nearchos, a seafarer of 325 B.C., called the mangroves "a forest growing in the sea" when he reached their northern limit in the Persian Gulf.

The mangrove forest contains only a few species of specialized Dicotyledonous trees. In Madagascar there are two kinds of mangrove forest: seacoast forest parallel to the beach, facing the sea, and estuary or bay forest in more brackish water and river mouths. Cabanis *et al.* estimate that the Malagasy mangroves cover 21,000 ha, or 1/570 of the island.

Mangrove needs particular conditions to survive: a shallow beach platform, high tide range, gentle waves, and an influx of brackish water. This last condition could not be satisfied by rainwater for much of the south, but there are underwater springs along the coast which drain the deeper layers inland. At Madiorano, 35 km north of Tulear, the villagers wait for low tide to go down the beach and collect fresh water from their spring.

Five main mangrove species grow in the southwest. They are easy to recognize, even at a distance.

Two of the species produce upward pointing rootlets in a 2 m radius round the tree. The root ends are called pneumatophores since they "breathe" oxygen for the plant. *Avicenna marina* (Avicenniaceae) has whitish undersides to its leaves which gives the whole tree a silvery look. Its pneumatophores are very thin. *Sonneratia alba* (Sonneratiaceae) has green leaves and thick, blunt-ended pneumatophores. Thus one can distinguish the pneumatophores even if they emerge from the mud side by side.

Three Rhizophoraceae, lacking pneumatophores, grow in the mangrove forest. They have viviparous fruits which grow a long root or radicle even while hanging on the tree. These pendant fruits have been given the common name "candle-trees". When the fruit detaches itself, it arrows into the mud, root foremost. Rootlets grow out of dots on the root, and if the fruit happens to fall awry the rootlets pull it upright.

Various characteristics distinguish the Rhizophoraceae from each other. *Rhizophora mucronata* has adventive buttress roots like piles which anchor it solidly in the mud. Stomata on these roots play a major role in the plants' respiration. The leaf terminates in a mucron, or small point. The fruit-embryo is very elongate. In *Bruguiera gymnorhiza*, instead of pneumatophores, knotty elbow-like protruberences emerge from the base of the trunk in a 1 – 2 m radius to anchor the tree in the sediment. The absence of mucron on the leaves or stilt roots help to distinguish this species from Rhizophora. Besides, the *Bruguiera* fruit has a persistent calyx in tongue-like strips, while the radicle is thick, canalized, and cigar-shaped.

In *Ceriops boviniana*, woody blades grow from the base of the trunk. These wing-shaped buttresses again brace the tree on the sand. The fruit-embryo, smaller than *Rhizophora*'s, has oval leaves rounded at the top.

Finally *Lumnitzera racemosa* (Combretaceae) is a borderline mangrove, growing in contact with terrestrial vegetation.

V. MEDICINAL PLANTS

Among the medicinal plants listed by Pernet a large number grow in the Malagasy southwest. In the Tulear region alone we have gathered 78 species in current use. Among the best known we could cite *Cedrelopsis grevei* the Katrafay, whose bark soaked in bathwater can relieve muscular fatigue. *Tamarindus indica* (Cesalpaceae), the Kily, bears bark which soothes rheumatism and measles, leaves good for urinary infections, and laxative fruits rich in vitamin C. Macerated leaves of *Uncarina stellulifera* (Pedialaceae), the farehatra, produce an anti-dandruff and anti-baldness shampoo. Different plants are used for diverse maladies: stomach aches, haemorrhage, wounds, tumours, eczema, venereal diseases, cough, fever, urinary infections, toothache, eye infections, rheumatism, and to aid convalescence after childbirth. A very high percentage of medicinal plants grow on the calcareous plateau and the red and white sands.

CONCLUSION

This rapid overview of the flora of southwest Madagascar is striking in the high level of endemism. Koechlin has estimated 18 endemic genera in this single region. Natural selection here is severe: only plants which can make do with very little water have survived. They survive thanks to diverse adaptations: overdeveloped root systems, enlarged, succulent trunks and branches, leaves either fleshy or reduced to thorns or else even non-existent, and waxy or hairy coatings. These adaptations rediscovered by plants of different families lead to strong convergences of form. From now on anyone who says "bottle tree" should not only think of baobabs (Bombaceae), but also of the elephants' foot *Pachypodium* (Apocynaceae), of *Moringa* (Moringaceae), of *Delonyx* (Cesalpiniaceae), of *Commiphora*, in the Burseraceae, and many others besides.

The oddest conclusion is to realize that the southwestern bush, at first so unenticing with its sad grey colour and its defensive spines, jealously guards so many medicinal plants which are daily used and greatly prized.

REFERENCES

Basse, E. (1934) Les groupements vegetaux du Sud-Ouest de Madagascar. Masson et Cie, Paris.

Cabinis, Y., Chabouis, L. and Chabouis, F. (1970) Vegtaux et groupements vegetaux de Madagascar et des Mascareignes, tome II BDPA, Tananarive.

Costantin, J. & Bois, D. (1907) La vegetation du Sud-Ouest de Madagascar. *Ann. Sc. Nat.*, 10° ser., Bot., 16, 95 – 225.

Debray, M., Jacquemin, H. and Razafindrambao, R. (1971) Contribution a l'inventaire des plantes medicinales de Madagascar — Travaux et documents de l'O.R.S.T.O.M., n° 8.

Humbert, H. (1927) Principaux aspects de la vegetation a Madagascar. *Mem. Acad. Malg.*, 5, 1 – 89.

Humbert, H. (1954) Les territoires phytogeographiques de Madagascar. *Coll. Reg. Ecol. Globe, CNRS*, pp.191 – 204.

Keraudren, M. (1961a) Quelques aspects des formations xerophiles du Sud de la Republique Malgache. *Bull. Soc. Bot. Fr.*, 108, 73 – 79.

Keraudren, M. (1961b) Au pays des arbres serpents. *Sc. et Nat.*, 45, 3 – 8.

Keraudren, M. (1963) Pachypodes et Baobabs a Madagascar. *Sc. et Nat.*, 55, 2 – 11.

Keraudren, M. (1966) Types biologiques et types de succulence chez quelques vegetaux des "fourres" du Sud-Ouest de Madagascar. *Mem. Soc. Bot. Fr.*, 157 – 163.

Koechlin, J., Guillaumet, J.L. and Morat, P. (1974) Flore et vegetation de Madagascar, (J. Cramer, ed.).

Pernet, R. (1957) Les plantes medicinales malgaches. Catalogue de nos connaissances chimiques et pharmacologiques, tome VIII, serie B.

Pernet, R. (1959) Les plantes medicinales malgaches. Memoires de l'Institut Scientifique de Madagascar, tome IX, serie B.

Perrier de la Bathie, H. (1921) La vegetation malgache. *Ann. Mus. Colon. Marseille*, 3° serie, 9, 1 – 268.

Perrier de la Bathie, H. (1936) Biogeographie des plantes de Madagascar, Paris.

Poisson, H. (1912) Recherches sur la flore meridionale de Madagascar, these Paris.

Poisson, H. (1921 – 22) Monographie de la province de Tulear. *Bull. Econ. Madagascar*, 18, n° 3, 37 – 65; n° 4, 43 – 73.

Rabesandratana, R., Rakotozafy, A. and Thomasson, M. (1977) Le fourre des dunes de sables blancs dans les environs de Tulear (Sud-Ouest Malgache). *Ann. Univ. Mad., serie Sc. Nat. et Maths*, n° 13, pp.117 – 30.

Rabesandratana, R. (1977) Resultats d'enquete et de localisations de plantes medicinales de la region de Tulear. *Ann. Univ. Mad., serie Sc. Nat. et Maths.*, n° 13, pp.131 – 50.

Rabesandratana, R., Rakotozafy, A. and Thomasson, M. (1977) Approche floristique et ecologique de la vegetation des environs de Tulear (Sud-Oest malgache). *Ann. Univ. Med. ser. Sc. Nat. et Maths*, n° 14, pp.205 – 22.

Thomasson, M. (1974) Essai sur la physionomie de la vegetation des environs de Tulear (Sud-Ouest malgache). *Bull. Mus. Nat. Hist. Nat.*, 3° ser., 250. Ecologie generale 22, 1 – 27.

CHAPTER 4

The Invertebrates

PAUL GRIVEAUD

In the Animal Kingdom the invertebrates number more than a million species, divided into 116 orders and 47 classes, while the vertebrates are limited to a few tens of thousands of species, 45 orders, 8 classes. It is almost impossible to estimate the numbers for Madagascar. In spite of the enormous amount of scientific work already accomplished, the world of the invertebrates is still largely unstudied and undiscovered. The island holds, perhaps, more than 100,000 invertebrate species. In the butterflies alone, which number among the best-known Malagasy invertebrates, we have found well over 3000 species. It is obviously out of the question to discuss all the invertebrates here. We shall only point out a few which the traveller or naturalist might encounter in the course of his peregrinations about the island. We shall also leave aside Protozoa and marine forms, and choose our examples among the worms, gastropods, myriapods, crustaceans, arachnids and insects.

WORMS AND GASTROPODS

Among the astonishing species of the Malagasy fauna, we shall start with the huge flatworms (15 cm and more), coloured in brilliant contrasts of black and red, which wriggle across the soil of the Eastern rainforests.

Little forest leeches (''dinta'') in Malagasy, live on the soil and leaves of these same forests. They make exploration unpleasant, particularly during rains. They are very thin and difficult to see, and can slide through cracks of shoes and clothing to fasten on one's skin and bloat with blood.

There are about 37 genera of Malagasy terrestrial molluscs, 7 of them endemic. Eleven are shared with India, 4 with both India and Africa, and only 6 with Africa alone. None have South American affinities. The commonest genus, *Tropidophora*, includes 85 of the island's approximately 300 species and extends to Europe and Arabia.

Terrestrial gastropods include large forms like *Ampelita*, once consumed in huge quantities by the Malagasy of Lake Alaotra. *Clavator* is a genus with numerous species which is also known as a fossil in Africa, and commonly used by geologists to date fossil layers. The large majority of the 300-odd species are endemic to the island. The regions richest in species are the calcareous regions of the north, west and south. Other species, mostly small and slug-like, haunt the eastern humid forests.

We should also point out that the giant snail, *Achatina* has been introduced from Africa by man. It is widespread in the east and the Sambirano where it attacks the fruit of cacao trees. It is brownish and conical, and may reach 5 – 6 cm shell length. People eat *Achatina* in Africa and the Pacific, but not in Madagascar.

SPIDERS AND SCORPIONS

Madagascar has about a dozen species of scorpions divided into 7 endemic genera. They have generally African affinities, but most Malagasy scorpions are small, not the enormous black species of Africa.

It remains true that they are disagreeable beasts with a painful or even dangerous sting. They live in both forest and savana. They are mainly nocturnal, with the unpleasant habit of taking refuge by day under any available cover, and are often found beneath the tent groundsheet when you strike camp. It is prudent to shake out your shoes in the morning before putting them on, particularly in the West and the southern bush where there are few hiding places.

More than 400 species of spiders, in 39 families, are known in Madagascar, though they are very incompletely studied as yet. We shall mention a few of the most obvious ones.

Nephila, and particularly the species *Nephila madagascariensis*, are large, long-legged spiders whose huge webs hang everywhere, even under the floor-timbers. Some time ago Reverend Father Camboue tried to start a silk industry using these spider webs, but the attempt failed. However, very beautiful traditional *lamba* were woven from spider silk for the Merina Kings in precolonial days.

The Gasteracanthea or crab spiders have about 20 Malagasy species in forests throughout the island. They are often brilliantly coloured, with abdomens ornamented with spines.

The Eusparassida are represented by a curious species, *Olios coenobita*, in the bush of the Mahafaly plateau. One is stupefied to discover certain thorn-bushes hung over with the whitened shells of little snails, dangling from threads as much as 80 cm above ground. The spiders have hoisted up empty shells by their silk, to form their own shelters. They manage to lift shells weighing 20 times their own weight!

The Arachaeidea are one of the oldest spider families of the world, originally described in many million years old Baltic amber. The genus *Archea* exists in Australia, South Africa, and Madagascar — a true Gondwanaland distribution. They are extremely small and difficult to see in the forest vegetation or the soil. They look like little monsters with outsized chelicerae and very long thin legs.

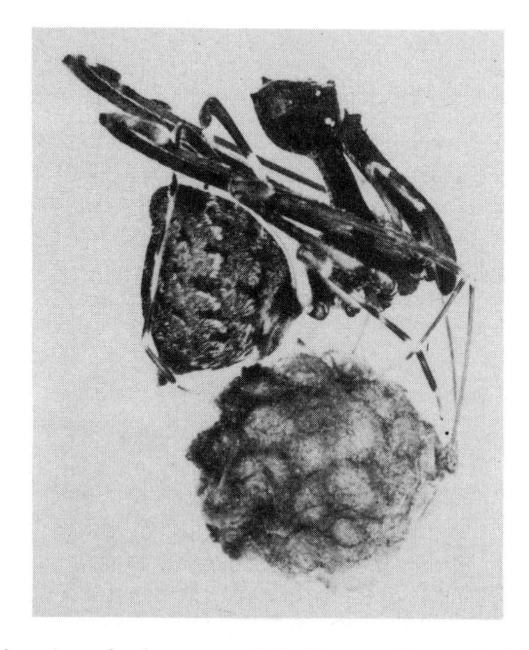

Fig. 4.1 *Archaea workmani* carrying her cocoon. This bizarre spider is only 3.5 mm long (Ph. Oberlé)

Among the sedentary spiders we should point out the aquatic species. Besides the freshwater *Argyronetes*, we know of two marine forms. They line the interior of little bell-shaped cavities with silk, either in the coral reefs or in clay nodules in the marine mud. They belong to two different species and genera: *Desis crosslandi*, (Baryche- lidae) and *Atrophonysia intertidalis*. Thanks to their "diving bells" which trap air through high tide, they can live at the low tide mark, where they make sorties in search of small invertebrate prey.

One venomous spider whose bite is actually dangerous is well known to Malagasy under the name "menavody" (*Latrodectus menavodi*). It is black, the size of a pea, with the end of the abdomen marked with one or more bright red spots. It hides in shadowy corners. The family is widespread: it's representatives in Europe and America are the "black widows".

Madagascar also has several small mygales, the tarantula group of spiders.

CRUSTACEANS

The truly terrestrial crustaceans of Madagascar are limited to isopods (wood-lice or sow-bugs). Crayfish live in the freshwaters, rather like those of Europe in appearance but in fact related to Australian and South American forms (Astacoides). There are also freshwater crabs (Potamonides) and shrimp. Some of the shrimp are small Atyidae and known as "patsa". They include a whole series of remarkable cave forms. Others are larger, like prawns, and much sought after by gourmets.

Among the oddities of the crustacean realm we might point out a curious little amphipod, *Austroniphargus bryophilus*. It is confined to the specialized ecological niche of the thin layers of water on the thick moss of natural basins in the granite rocks of a few mountaintops, in particular, the Andringitra massif. A related species has been recently found in the springs near Fort Dauphin.

MILLIPEDES AND CENTIPEDES

These have many Malagasy species. The Chilopodes, or centipedes have flattened bodies with a single pair of legs per segment, and move extremely rapidly. There are two large *Scolopendra* species, introduced by man and common throughout the tropics. They give a painful bite.

The Diplopodes, or millipedes have a rounded body, two pairs of legs per segment, and move more slowly. They are particularly frequent in the humid forests. *Sphaerotherium* are short and wide, brilliant green or matte brown. The Iules are thin and can grow to 15 cm long. They are black, though with red and orange variants. Millipedes are inoffensive vegetarians in spite of excreting a red liquid when disturbed. They roll up in a hard ball (*Sphaerotherium*) or a flat spiral disc (Iules) to protect themselves.

INSECTS

With the prodigious world of insects we enter an immense domain. In Madagascar their species number tens of thousands. We can only cite a few examples chosen among each of the Malagasy orders.

Although most Malagasy mammals and some of the birds belong to endemic families, it is not the same for insects. This ancient group has almost no endemic families. On the other hand there is a very high degree of endemism at the level of genus and species.

The insects of Madagascar are neither larger nor differently coloured from those of other lands, but the varied terrain, climate and vegetation have allowed the few forms which first occupied the island to diversify almost infinitely. Madagascar has acted like a natural laboratory. The result of this isolated evolution is of the greatest interest to biologists, as well as for the ecologists who study natural equilibria. Perhaps no other region of the globe has shown such active speciation and adaptation to every habitat.

Let us correct, in passing, a common error. Entomology, the study of insects, is not just a theoretical occupation or a dreamer's distraction. Entomology results in highly concrete applications. It distinguishes useful from harmful insects, and provides means of combatting the latter. Some billions of billions of plant-eating insects live on the earth's surface, causing agriculture to lose the equivalent of the budget of a large nation every year. Without the long, patient research of entomologists, these depradations might be multiplied a hundredfold. We should also remember entomology's contribution to medicine, for instance by discovering the role of the Anopheles mosquito, and thus beginning to free mankind from the scourge of malaria.

From dragonflies to crickets

The Odonata or dragonflies and their two suborders, Anisoptera and Zygoptera, are represented in Madagascar by 10 families, 52 genera of which 13 are endemic, and 148 known species of which 101 are endemic. The largest species reach 13 to 14 cm wingspread,and are widely distributed in Africa and Madagascar (genus *Anax*). Many species of graceful damselflies gleam blue or red, along the forest streams.

The Dictyoptera include some 100 species of cockroaches of which 96 are endemic. There are about 60 mantids, some of them astonishing leaf mimics like the females of the genus *Brancksikia*.

Isoptera, or the termites, are represented by 17 genera and about 75 species, 71 of them endemic. We can divide them into three groups: termites which attack wooden buildings, termites attacking cultivated plants and live trees, and soil termites. Only the last group are particularly noticeable to travellers, particularly in the open savannahs of the west and south where their hills form small

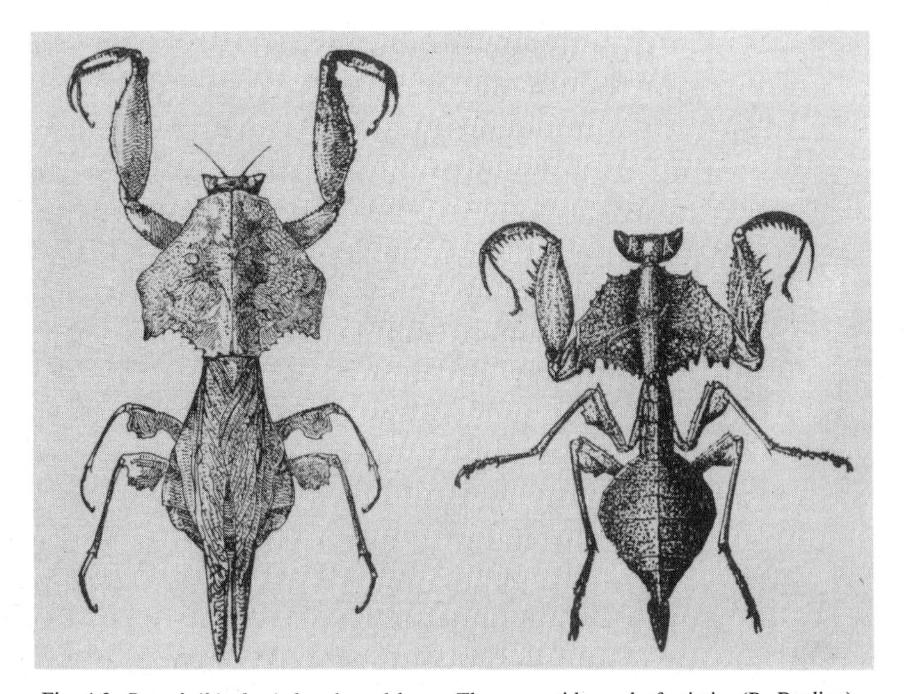

Fig. 4.2 *Brancksikia freyi*, female and larva. These mantids are leaf mimics (R. Paulian)

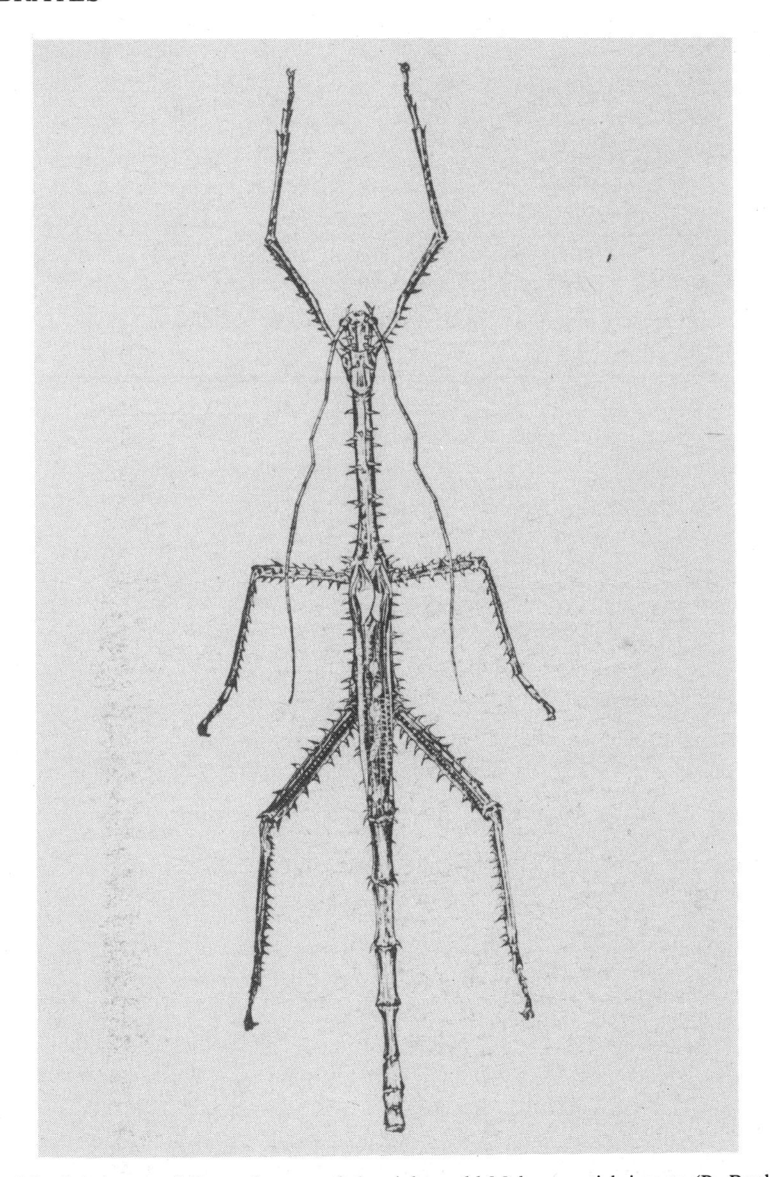

Fig. 4.3 *Achrioptera fallax* male, one of the eighty-odd Malagasy stick insects (R. Paulian)

characteristic cones. No Malagasy species build the monumental termite mounds several metres high which one sees in Africa.

The Cheleutopteas or phasmids have about 80 Malagasy species, all endemic. Some of the large phasmids of the genus *Achrioptera* mimic bunches of thorny twigs so cleverly that they are very difficult to see in spite of their bright colours. The largest females measure 25 cm long. Their short wings, which scarcely allow them to fly, are chiefly used in defence. The insect spreads them in ghostly, frightening fashion — but most of the time a frightened phasmid is immobile and escapes notice.

The Orthoptera, (grasshoppers, locusts, crickets) include a very large number of endemic species. The migratory locusts of the genera *Nomadis* and *Locusta*, under certain climatic conditions pullullate and become gregarious. They then move in immense flights of millions of individuals, descending to ravage

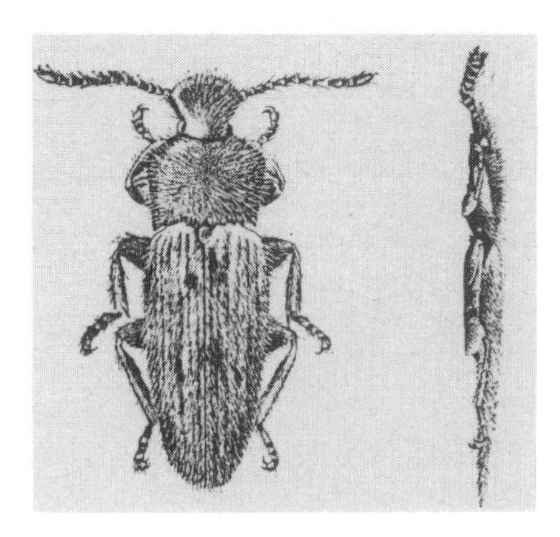

Fig. 4.4 *Tabula depressissima* dorsal and side views. This is an Elaterid beetle, flattened for living in the leaf axils of *Ravenala* the Travellers' Palm (R. Paulian)

the crops, and laying in particular spots. The eggs produce further armies of wingless nymphs which are as destructive as the adults. There was a time when these flights arising from the Southwest where the species reproduces, reached the High Plateau, and descended in clouds on the island's capital.

An anti-locust service was created, with a large centre at Betioky. Efficient controls have reduced the swarms, but the least relaxation of control could lead to renewed swarming. The rural people of the west and south do not disdain the nymphs as food, as well as some of the large crickets of the west which dig deep tunnels in the earth.

Two Acridian species of the genus *Phymateus*, "Valalan'alika", are remarkable for their garish colour: elytra steel blue with yellow checkerboard, and red wings, or else elytra brown to olive green and orange wings. The thorax bristles with rough tubercles.

Beetles

The Coleoptera are one of the best studied orders to date, with a large number of described species and genera. We shall only mention families with particularly remarkable species.

The Elateridae have large, beautiful species in the western dry forests and southern bush. They reach 45 mm long, with white or cream elytra decorated with black patterns.

The Buprestidae are one of the large Malagasy families of Coleoptera. More than 500 species are known, with practically 100% endemism, and the study of the small species of the forest canopy is far from complete. This swarming universe of insects includes many superb *Polybothris* species, metallic green or blue. We know of 160 species of Coccinellidae, a third of them endemic.

The Chrysomelidae number nearly 800 species, nearly all endemic. Many have handsome metallic colours. The most remarkable belong to the subfamily Sagrinae, with curiously swollen tibia on the hind legs.

The Anthribae have black species with white patterns. Their short rostrum gives them somewhat the appearance of weevils.

The Cerambicidae or long-horned beetles have 600 described species, nearly all endemic. The largest belong to the subfamily Prioninae and to the genus *Hoplideres*, where some females reach 85 mm in length, not counting the long antennae. These Prioninae are mono- chrome insects ranging from dark brown to light brown. They are nocturnal, and generally develop in dead wood.

The subfamily Lamiinae, though without such large species, has prettier ones. Some are diurnal, brightly coloured, with antennae which can be twice the length of the body. One of the commonest is *Stellognatha maculata*, whose black body has spots of pure white.

The Brenthidae are curious beetles with straight, elongate body, variable size and usually strong sexual dimorphism, the male often with neck and rostrum disproportionately elongated. Seventy-one species have been described, seventy of them endemic, divided into 39 genera of which 23 are endemic.

The Circulionidae or weevils have an enormous complex of families. Some 1300 species have already been described, only 3 or 4 non-endemic! Forms and sizes are extremely variable. Some mimic bark, moss or lichens. The strange group of the Attelabes roll up leaves into long cigars, to shelter an egg and the larva which emerges. Males often have head and thorax prodigiously elongated, which has earned one species the name "Giraffe". In the Tananarive region one species of this group attacks cultivated beans.

The Lucanidae are only represented by a few small species that are wholly endemic.

The Scarabeidae include a huge number of genera and species unique to Madagascar. Among the Scarabeinae (dung beetles) we might mention the remarkable group of forest-living *Eplissus* with their brilliant metallic colours, and the *Scarabaeus radama* and *Neonematicum sevoistra*, common in the southwest and the south, which you often meet on earthen paths rolling along their balls of dung destined to feed the larvae. The Melolonthinae or cockchafers again have many species and genera, of which the oddest are large pure white chafers. One of the most spectacular subfamilies is the Cetoniinae or flower beetles, with about 280 species divided into about 50 genera. The Malagasy flower beetles have many pretty species of various sizes, often with bright metallic colours. Among the loveliest are the lustrous and velvety *Euchroea*, like *E. coelestis* and *E. urania* whose larvae develop in *Pandanus* axils.

Among the Carabidae, carnivorous beetles, let us mention the Cicindeles, where the *Pogonostoma* species have South American affinities. They are arboreal with an elongate, cylindrical body, and steel blue or bronze-green colour. A large metallic blue Carabid, *Ctenostoma bastardi*, lives in the southwest. The Scaratini, redoubtable digging carnivores with large mandibles, are represented by about 60 species in 26 genera.

The order of Planipennidae includes, in the family Myrmeleonidae or ant-beetles, 17 described species. The Ascalaphidae have large forms in the west and south, including *Palpares voeltzkowi* with violet-black and white wings, and reddish-brown abdomen.

Butterflies and Moths

The Malagasy Lepidoptera, like the Coleoptera, are relatively well-known. Their often brilliant coloration and their great variety have attracted amateurs as well as professionals, so they have been studied in the field as well as the laboratory. The caterpillars of many species, though not the most spectacular ones, have become serious crop and storage pests.

On the High Plateau with its temperate climate you mainly find smallish, discretely coloured butterflies and moths. Leaving the capital, the amateur will find more beautiful species along forest roads in the Eastern escarpment, or on the beaches of damp sand by rivers and streams between Moramanga and Anosibe Anala, or around Prinet. But it is in the coastal regions that life explodes with exuberance, with magnificent large species in amazing colours.

Several Malagasy Lepidoptera have an international reputation, and are much sought by collectors. Some of these fortunately, are common like *Chrysiridia* (Urania) and certain *Papilio*. Others are rare or highly localized. Commercial collecting in habitats which are themselves threatened with destruction, may lead to extinction. This is true of the comet, *Argema mittrei*, and certain *Charaxes* and *Papilio*. Sale and exportation of these species should be controlled, and authorized only for specimens from hatcheries under supervision.

Apropos of the sale of butterflies, we might point out that the glass cases sold by souvenir merchants in Tananarive often contain some species which are not Malagasy, but imported from Guyana or Brazil, such as the Morphos, huge butterflies with brilliant blue reflecting wings.

Some 64 families of Lepidoptera occur in Madagascar, with a whole range of shapes, colours and sizes from Microlepidoptera to Macrolepidoptera. It is impossible to sum up the number of genera and species: not only are the original descriptions scattered through many publications, but whenever a family is revised, new genera and species appear. Just for example, the Lymantriidae (ex Liparidae), revised in 1977, included before that date 24 genera and 182 known species. The 1977 revision added 28 new genera and 86 new species! Furthermore, at the moment only 13 of the 64 existing families have had a recent revision! Nonetheless, just to put forward a figure, we estimate more than 3000 species of Lepidoptera have been described, of which 97 – 98% are endemic.

It is strange that Madagascar has no species of Homoneura. These are primitive Lepidoptera, best known by the lovely species of the family Hepilidae or ghost-moths. In Africa and Australia, they reach a wingspan of over 20 cm, their brightly coloured wings spotted with pale mother-of-pearl.

Caterpillars of the family Psychidae live inside sheaths covered with silk or bristling with twigs. *Deborrea malagassa*, the ''fangotabolo'', is very common. On the High Plateaux it eats mimosas, and you often see the grey larval sheaths, as big as a hazelnut, hanging from the branches.

Among the Geometroidea there are beautiful species, particularly among the Uraniidae, a family that includes both delicate white forms and the well known, *Chrysiridia madagascariensis* or Urania moth. This gorgeous insect has a palette of iridescent greens and oranges on a background of velvet black. It migrates across the island in huge groups.

The superfamily of the Noctuoidia has many Malagasy representatives in its families the Noctuidae, Arctiidae and Amatidae.

In the Noctuidae there are large species in the subfamilies Ophiderinae and Catocalinae. In the latter the genera *Miniodes* and *Miniophyllodes* have forewings of yellow ochre and hindwings carmine. The genera *Cyligraramma* and *Erebus* are large nocturnal, slow flying, moths which take refuge in dark places by day and frequently enter habitations. Malagasy legends say these are ''lolo-paty'' come to visit the living. ''Lolo'' in Malagasy means both spirit and moth; ''lolo-paty'' can be translated ''spirits of the dead''.

The Catocalinae also has large, handsome species in the genus *Emmonodia*.

The family Agaristidae has 35 species and 11 genera in the Island. These pretty moths fly by day in the forests where it is difficult either to observe or capture them. The largest and handsomest are *Rothia*, a genus with 23 species. The forewings have red, white or blue spots on a velvet black background. The hindwings are variable, often with a wide black border.

The most amazing moth of this family is *Pemphigostola synemonistis*. The male has a hyaline stridulatory apparatus with narrow transversal ridges at the base of the leading edge of its fore-wings. It is the only Malagasy lepidopteran with a stridulatory organ, which is unusual for any insect of this order. The sexes are fairly dimorphic and the female lacks this organ, whose role and use are so far unknown.

The large, mainly nocturnal family Arctiidae has very small moths in the subfamilies Lithosiinae and Nolinae, and a few pretty, medium sized species among the Actiinae and Aganainae. Many of their caterpillars live among lichens and mosses.

The small family Amatidae has about 80 species in 16 genera, of which 15 are endemic. These are small moths, difficult to observe, brownish or black with white or yellow patterns. Only the genus *Euchromia*, also found in Africa and the Comoros, has two brighter species which reach 50 – 55 mm wingspan.

The superfamily Bombycoidea or silkmoths, has just one species of the family Eupterotidae and one of Bombycidae. The best represented families of this group are the Attacidae or Saturnidae, and the Lasiocampidae.

Among the Attacidae, the best known is *Argema mittrei*, the comet moth, largest of the Malagasy Lepidoptera and one of the largest in the world. The male can exceed 20 cm length including tail. *Argema* is a clear yellow, with a few russet marks, distinguished by hindwings prolonged as two long "tails" which are especially long and narrow in the male. This nocturnal species is not rare, far from it. However it is sold commercially to such a degree that it could easily become rare. The Malagasy gatherers can easily recognize the 6 – 10 cm cocoon of white or pinkish silk, and gather hundreds to obtain fresh, intact adults at the moment of emergence. The caterpillars eat many sorts of plant and could easily be reared. It would be preferable to protect this species in the wild and have specialized hatcheries offering specimens for sale.

Still in the Attacidae, the genus *Tagoropsis* has fine large forms, in neutral yellow or brown colours.

The family Lasiocampidae in its two subfamilies contains almost a hundred Malagasy species in 27 genera. These nocturnal moths, coloured dull brown, reddish, or greyish, seem to have little attraction except for the specialist. However the Malagasy have been using *Brocera* silk for well over a century. They unrolled the cocoons to form a silk which was fairly coarse and raw but very solid. It was woven by the women, especially in the Betsileo country. This silk, in former times was exclusively made into "lambamena", the magnificent red and black dyed shrouds of the dead. Attempts to raise the caterpillars have been unsuccessful and cocoons were simply gathered from the wild for local craftsmen and women. With the introduction of the silkworm (true *Bombyx*) and the importation of silk thread, the use of *Brocera* cocoons has largely disappeared since they are somewhat prickly with hairs from the caterpillar. The Malagasy call *Brocera* caterpillars, cocoons and moths "Landy" or "Landibe".

The Sphingidae or sphinx moth family has about 60 Malagasy species in 24 genera. They are powerful fliers and there are several species of world-wide distribution, but there are also no less than 34 endemic species and 8 endemic subspecies. One of the most interesting Malagasy sphinx moths is *Xanthopan morgani praedicta*, the only species which fertilizes the comet orchid, *Angareceum sesquipedale* (see Chapter 2). It is one of the largest of the world's Sphingidae with a wingspan of 15 cm. Another sphingid, *Euchloron megaera lacorderei* is a handsome insect with bright green forewings and yellow hindwings speckled with black.

Among the diurnal butterflies, the Rhopalocera, we find most of the brightly coloured species.

The family Papilionidae is represented in Madagascar by only 13 species in 3 genera. *Graphium* has 3 endemic species of small wingspan. In *Papilio*, the swallowtails, the *P. demodocus* group has two endemic species: *P. grosemithi* and the rare *P. morondavana*. The *P. dardanus* group has one endemic subspecies, *P. dardanus meriones*. The *P. nireus* group has four endemics: *P. oribazus*, *P. epiphorbas*, *P. delandei* and *P. mangoura*. Finally the genus *Pharmacophagus* has one beautiful endemic species, *P. antenor*, which lives in the west and Southwest, a huge black butterfly with white markings and red lunules on the hindwings which are prolonged into the swallowtails.

The family Pieridae are butterflies with white wings marked in black or orange. They have eight genera and about 20 species. Only the genus *Milothrys* with large, rare species, contains pretty butterflies with orange wings 60 – 65 mm in wingspan.

The Danaidae have only two genera with one species each: *Danaida chryssipus* and *Amauris nossima*.

The Nymphalidae, with 18 genera and 45 species, include handsome butterflies much sought after by collectors. We may cite *Euxanthe madagascariensis*, seven species of *Charaxes* and four species of *Hypolimnas*, as well as the pretty little *Precis radama*, which is bright blue.

The Acraeidae, small butterflies (except the fine *Acraea hova*), have pretty forms with more or less translucent wings. They have 19 species in 2 genera.

Finally, the families Lycaenidae and Satyridae have many endemic species of small and medium sized butterflies, but monographs on these families have not yet been published.

From flies to water scorpions

The huge order Diptera, the flies, mosquitoes and so on, has more than 70 families present in Madagascar, with many genera and species, but it has not been the object of a summary publication. A few families are well studied such as the Blepharoceridae and the Pyrogotidae. One volume of the Faune de Madagascar deals with the Culicidae-Anophelinae mosquitoes.

Many of the flies are parasitic on man or animals. The Simuliidae have aquatic larvae, and a very disagreeable bite. These little flies the "mokafoy", pullullate in various regions. The Ceratopogonidae are tiny gnats which live by the edges of the sea and saline lakes. They are practically invisible but their bites are even more disagreeable than the Simulid's. The Anopheles are vectors of various diseases including malaria. Finally in the hot regions of the west and south several species of Tabanidae, or gadflies, also bite painfully.

Mammals and birds have Dipteran parasites, some of them wingless. Besides these bloodsucking Diptera, the Malagasy fauna has hundreds of other flies and gnats, including parasites of plants and of other insects.

With the Hymenoptera (wasps, bees, ants, etc.) we move to a large and economically important insect order. Many Hymenoptera play a primary role in maintaining biological equilibrium, as predators of other insects including the pest species of crops and forests. The abuse of chemical insecticides which destroy these Hymenoptera as well as the pests for which they are intended, can recoil on the user with even greater damage.

Although there are few studies in depth of the Malagasy Hymenoptera, we may estimate about 1700 known species, divided into 300 genera.

The Vespidae or wasps, the "fanemitra", with around 60 species, are all too easily found, particularly

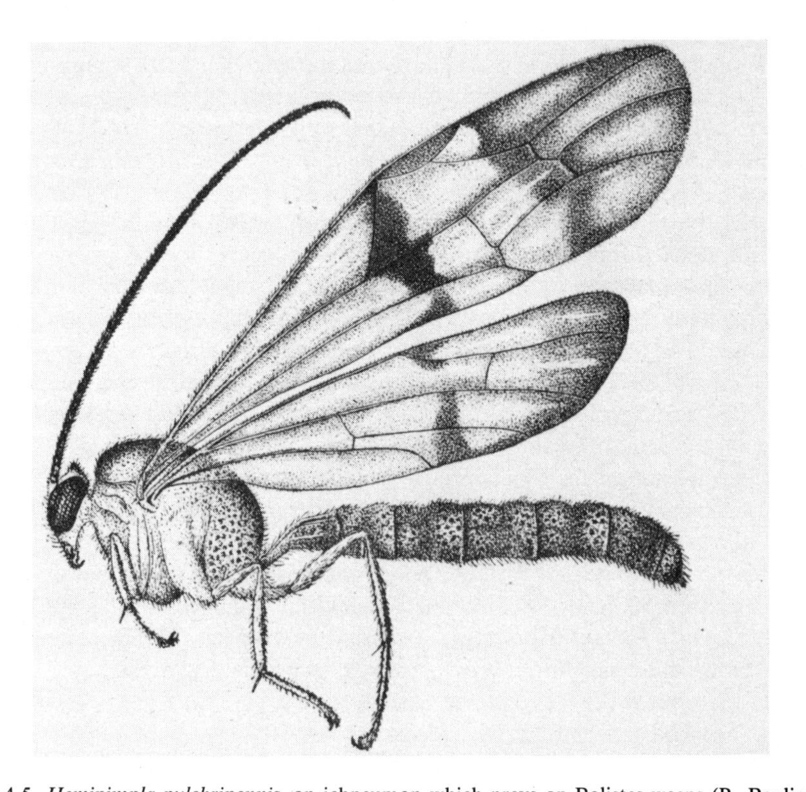

Fig. 4.5 *Hemipimpla pulchripennis*, an ichneumon which preys on Polistes wasps (R. Paulian)

in the dry western forests and bush. Their grey paper nests of open cells are hung from the vegetation at face or chest height, not very visible among the leaves. The inhabitants when disturbed become aggressive, and particularly the larger species have painful stings.

Malagasy bees give honey which is much appreciated, whether it comes from hives or honeycombs in the forest. Their Malagasy name, "renitantely", means "mother of honey".

The Formicidae or ants again have many species. On the forest trees you can see enormous balls more than half a meter in diameter which are the paper nests of tree-ants, genus *Crematogaster*. Covered galleries often connect these nests to the ground, running along the branches and down the trunk.

The Chrysidae are present, with 36 species that parasitize other insects, brilliantly coloured in metallic green. One can often see them flying round outside the woodwork of houses, looking for tunnels dug by the larvae of other insects, their prey.

The curious Mutilidae also have highly coloured forms. Their wingless females run on the ground; they are also parasites, particularly of caterpillars.

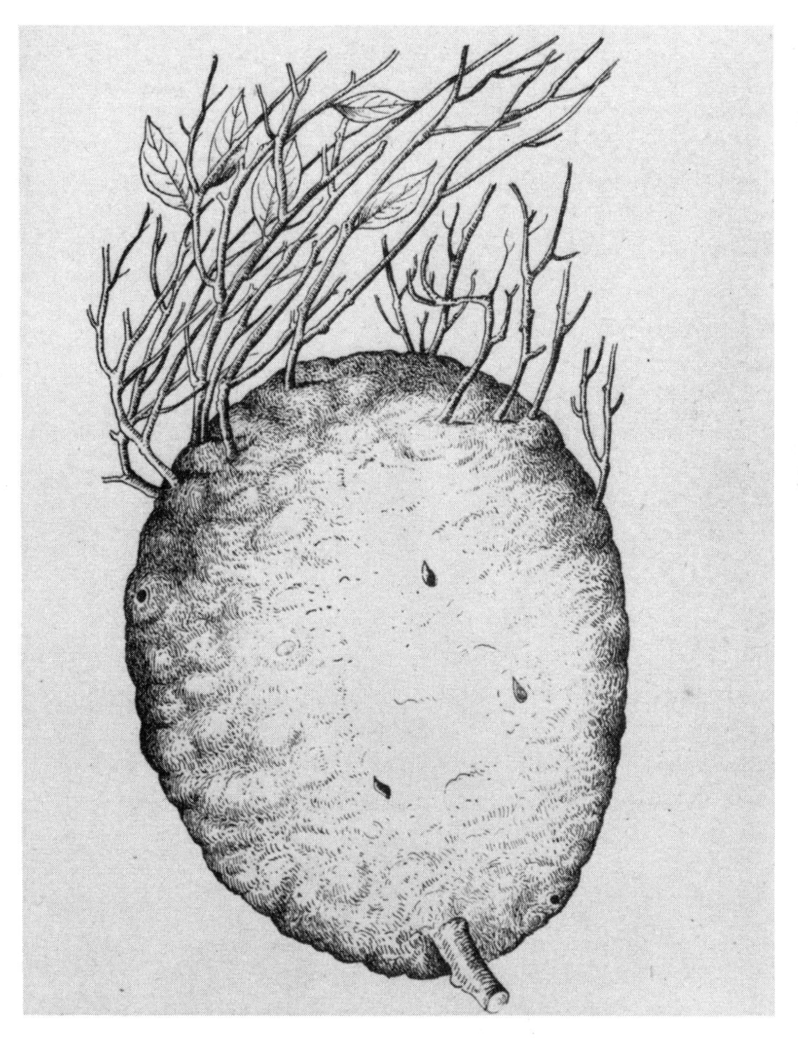

Fig. 4.6 The paper nest of arboreal *Camponotus* ants. These excrescences are common in both humid eastern and western deciduous forests (R. Paulian)

As a general rule the traveller should beware of all the Hymenoptera, and never catch them without precautions, for most inflict either bites or stings which are unpleasant or even dangerous.

The order Homoptera includes, among others, the Cicadidae or cicadas, the Yanidae or ciccadelles, the Fulgoridae, the Aphididae and the Coccidae or scale insects. Many families are still little studied, but an estimated 700 species are known.

About 30 cicada species in about 12 genera live in Madagascar. The large species of *Yanga* make the forests resound with their stridulation. *Yanga guttulata* may swarm, and its larvae become serious pests of sugarcane fields.

One common ciccadelle *Ptyelus goudoti*, dirty yellow coloured with black spots, attacks many legumes. In the parks and gardens of Tananarive, jacaranda trees frequently "rain" drops of foamy liquid as the larvae burrow into the host plant. This is a particularly common phenomenon in the jacarandas bordering Lake Anosy.

Among the Fulgoridae one common species, *Pyrops madagascariensis*, the "sakondro", is eaten by some groups of people. They pull off the rostrum, legs and wings, and serve the insects with rice either boiled or fried in oil. Many Fulgoridae and Flattidae (a related family with laterally flattened body, whitish or rose-red) live colonially as flower-mimics on trunks or branches. They secrete sugary manna, sap transformed as it passes through the intestine, which nourishes both ants and lemurs.

Among the scale-insects we must mention the "lokombitsika", a species of *Gascardia* which lives in fist-sized colonies, whitish, fixed onto branches. Each insect is enveloped in a thick layer of wax. These colonies attract ants, hence their name, "ant-wax".

The order of Heteroptera, the bedbugs etc., has many representatives. The families Nepidae, or water scorpions, and Belostomidae inhabit fresh water. The genus *Belostoma* includes a water bug which can reach 8 – 10 cm in length. These insects are attracted by lights in the evening, and often fall under streetlights or on house verandahs.

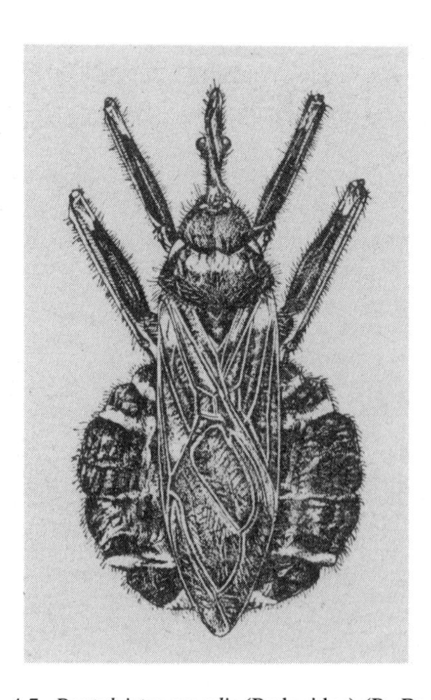

Fig. 4.7 *Pantoleistes grandis* (Reduvidae) (R. Paulian)

Fig. 4.8 *Gromphadorhina portentosa* a 60 gm cockroach which hisses in courtship or when alarmed (J. Fraser)

Fig. 4.9 *G. hopardi*. This cockroach also hisses when alarmed, but whistles in courtship (J. Fraser)

Among the Hemiptera only three families have yet been systematically revised, among about 15 families on the island. Some of these have many species, like the Coreidae and the Pentatomidae.

Many Hemiptera are serious crop parasites. Many are also attracted to lights, and the swarms of some Coreidae can be very disagreeable in the evening. One should be careful of the prettily coloured species of Reduviidae, which bite painfully. This family also contains remarkable species with thin elongate bodies and frail, outsized limbs, of which some live in caves.

Thus we finish our rapid survey of the world of malagasy invertebrates. It is a world with a prodigious number of species. These few pages can only show a pale reflection of its richness and endemism. However, it is a world under threat.

The invertebrates are fragile creatures, vulnerable to pesticides and insecticides, and to change in the natural environment with the degradation of forest cover.

Let us refrain from unthinkingly letting this fauna disappear. In spite of the small size of its individuals, it plays a primary role in the biological cycles of nature, and it offers an inexhaustible field for research.

BIBLIOGRAPHY

There are no general publications on Malagasy invertebrates. Information is dispersed through a large number of French and foreign books and journals. We should mention the volumes of the Grandidier collection, and the 38 volumes of the *Faune de Madagascar* which are consecrated to invertebrates. Among the periodicals, one should consult the *Bulletin de la Societe Entomologique de France*. Three chapters of Battistini, R. and Richard Vindard, G. eds., *Biogeography and Ecology of Madagascar*, (Junk, The Hague, 1972) deal with invertebrates. Also:

Paulian R 1950 *Insectes utiles et nuisibles de la region de Tananarive*, I.R.S., Tananarive, 120 pp. fig.

Paulian R 1951 *Papillons communs de Madagascar*, I.R.S., Tananarive, 90 pp. fig.

Paulian R 1954 Cinquante annees d'entomologie generale a Madagascar, *Bull. Acad. Malg.*, numero special du Cinquantenaire, 65 – 70.

Viette P 1962 Historique de l'etude des Lepidopteres de Madagascar, in *Noctuelles trifides de Madagascar*, These pour le grade de Docteur es Sciences naturelles, *Annales de la Societe entomologique de France*, t.131, fasc.I.

CHAPTER 5

The Amphibians

ROSE M. A. BLOMMERS-SCHLÖSSER and LEO H. M. BLOMMERS

1. INTRODUCTION

Neither Bufonids, nor caecilians seem to live on Madagascar. Frogs, Sahona or Bakaka in vernacular language, are well represented. About 150 species are known at present, all except two unique to this island. This number compares very well with continental faunas. For example from southern Africa about 100 frog species are known. Like the other vertebrate groups, the Malagasy frogs belong to only a few families, of which they represent a particular off-shoot. Eighteen of the twenty-two genera are endemic. Apparently, the Malagasy frog fauna represents the result of a vast radiation from a few ancient stocks.

There is still much to discover about these frogs. Until very recently, little more was known than the taxonomic accounts, based on mostly haphazard collecting. Guibe (1978) summarized this knowledge. As nearly forty new species have been described since 1970, it needs no stressing that information about life and ways of these frogs is still much wanting and that some earlier work in this respect is invalidated by misidentifications.

Arnoult (1966) published the first account of natural history of a Malagasy frog, *Mantella aurantiaca*. Arnoult and Razarihelisoa (1966, 1967) and Razarihelisoa (1973, 1974a&b, 1979) paid attention to about 20 species and Blommers-Schlösser (1975a&b, 1979a&b, 1982) described the larvae of about 50 species. The present account is based mainly on personal observations made between 1970 and 1973.

2. TAXONOMIC CLASSIFICATION

The present taxonomic survey is based mainly on Guibe (1978) with some modifications by the present author.

The family Microhylidae is represented by three endemic subfamilies, the Scaphiophryninae, the Dyscophinae, the Cophylinae and the widespread subfamily Microhylinae. Scaphiophryninae: *Pseudohemisus* (4 sp.), *Scaphiophryne* (1 sp.). Dyscophinae: *Dyscophus* (4 sp.). Cophylinae: *Mantipus* (6 sp.), *Anodonthyla* (3 sp.), *Stumpffia* (4 sp.), *Platypelis* (includes *Platyhyla,* 9 sp.), *Plethodontohyla* (7 sp.), *Rhombophryne* (1 sp.), *Madacassophryne* (1 sp.), *Cophyla* (1 sp.), *Paracophyla* (1 sp.). Microhylinae: *Microhyla* (1 sp.).

The family Ranidae is represented by the endemic subfamily Mantellinae and the widespread Raninae. Mantellinae: *Mantidactylus* (includes *Gephyromantis,* 59 sp.), *Mantella* (4 sp.), *Laurentomantis* (formerly *Trachymantis,* 3 sp.). Raninae: *Tomopterna labrosa*, *Rana tigrina* (introduced from Asia), *Ptychadena mascareniensis* (African species).

The family Rhacophoridae is represented by one endemic genus, *Boophis* (29 sp. = Rhacophorus). The monospecific genus *Aglyptodactylus* is close to the Rhacophoridae.

The family Hyperoliidae is represented by one endemic genus *Heterixalus* (8 sp.). The occurrence of the genus *Hyperolius* remains to be demonstrated (Blommers-Schlösser, 1981).

Our present knowledge of the life histories of Malagasy frogs is summarised in Table I. It is easily seen that adaptations to many different microhabitats are found. Especially, the Mantellinae and the Microhylidae show a vast adaptive radiation. It is conspicuous that most frogs are treefrogs, whereas the burrowing frogs represent only a small minority. Most species dwell in the forest. The savannah, which arose from clearing of the forest, is inhabited by relatively few species.

3. THE EASTERN RAINFOREST

The original vegetation of the eastern part of Madagascar is rainforest. The trade winds coming roughly from the North-East, bring clouds from the Indian Ocean, most of which precipitate on the eastern windward side, which rises sharply in great scarped steps. The island receives most of the rain in the southern summer from November to April. On the east coast no real dry season exists and the short rainless midwinter spell on the eastern slopes of the central highlands is mitigated largely by fog and dew. Almost needless to say most frog species live in the eastern forests.

Frogs are the least studied group of Malagasy vertebrates. As most species are known from one or two places only, a survey of their ecological properties and requirements is altogether impossible. Fortunately, however, there exist a few rather accessible places in the rainforest, which permits at least a local view onto the various ways Malagasy frogs live. One of these is Perinet, now called also Andasibe (not to be confused with many other villages of the same name). Originally named after a French engineer who died during the construction of the railway from Tananarive to Brickaville, this small village lies halfway down the eastern escarpment at approximately 900 m altitude. It is still surrounded by large areas of original, though often degraded forest. Between the hills water collects in all sorts of brooks and puddles. Here, more than fifty frog-species live, each taking its own place in an immensely varied biotope. A picture of the variety in form and habits of such a local fauna might give an impression of the uniqueness of the Malagasy amphibians.

Frogs are ecologically ambiguous, the requirements of larvae and adults being essentially different. Usually, the larval habitat can be easily defined, whereas the wanderings of the adults are often a matter of guesswork. Thus, in the following account the whereabouts of the tadpoles are considered in the first place.

3.1 Running water

In the rainforest you might find many kinds of running water. In the large rivers you cannot find tadpoles. Firstly, I will tell you about the clear forest brooks, which have a stony bottom and are provided with waterplants. Secondly, I will tell you about very shallow, slowly-running streams. The water is acid in the rainforest.

A search along forest books will usually reveal few frogs in daytime. *Mantidactylus femoralis* (Fig. 5.1) and *M. lugubris* are about the only species regularly met. They look like ordinary frogs and jump very well. The enlarged discs on fingers and toes provide a sure foothold on the slippery stones, with which their black and brown coloration is hardly in contrast. When disturbed they dive, even in fast running streams and are never seen again. *M. opiparis*, another ground dwelling species, also seems to prefer the banks of brooks. Like the other species, it is active during the day and early evening all times

Fig. 5.1 *Mantidactylus femoralis*, male

Fig. 5.2 *M. grandidieri*

of the year. When threatened it disappears into the vegetation. We never saw it jump into water. *M. opiparis* is not a very good jumper.

More mysterious is the life of two giants among the Malagasy frogs, *M. guttulatus* and *M. grandidieri* (Fig. 5.2). Both can reach a size of 10 cm and can be met on the banks of small streams, but only after sunset in the rainy season. Where they stay the rest of the time and where they breed is still unknown. Because they are offered for consumption on the Zoma (market) of Tananarive, the discovery of their whereabouts should not be too difficult. Most species which breed in streams are however treefrogs of the genus *Boophis*. To meet these, one must go out on a rainy summer evening, when they descend from the trees to congregate near their breeding sites. The males call in large choruses. Standing in the dripping darkness of the forest, one can hear many different sounds from high to low out of the trees. *Boophis luteus* (Fig. 5.3) is very common in this area and calls like the siren of an American police car. It is not easy to locate the different songs, but using a torch one can capture calling males. Females are rarely met, most of them with a male in amplexus. In Perinet, 11 out of 15 *Boophis* species were found near

Fig. 5.3 *Boophis luteus*, female

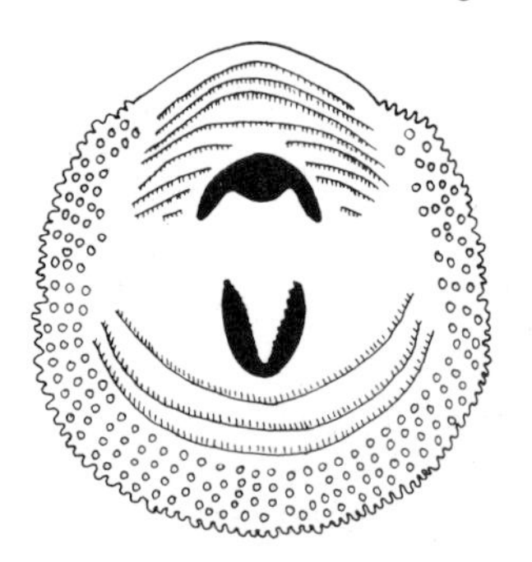

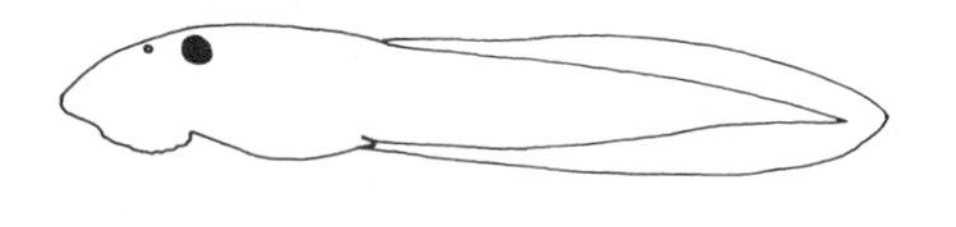

Figs. 5.4, 5.5 *Boophis erythrodactylus*, torrent tadpole, mouth and lateral view (L. Blommers)

Fig. 5.6 *Boophis erythrodactylus*, male (L. Blommers)

streams and only 4 close to pools. It is not unusual to find 5 species over a few hundred metres alongside a stream. The eggs of *Boophis* species are difficult to find. All those we saw were batches laid in the water, mainly attached to plants.

In contrast to the eerie deafening concerts on rainy summer evenings, only the rippling of water is heard on winter nights. Apparently all stream-breeding frogs around Perinet mate in the summer. Some species have a short breeding season of a few weeks, while in others mating takes place over some months. The development of the tadpoles is rather slow in these cool waters. As far as could be observed, metamorphosis in all species occurred in September and October just before the beginning of the next summer. The emerging young can then find enough small prey when the rains start and grow to a reasonable size before next winter. Remarkably, many young of the genus Boophis have a juvenile hue, which is predominantly green and probably serves for camouflage. Species of this genus reach a size of 2 – 10 cm as adult.

The winter is evidently the best time to search for tadpoles in the forest brooks, since the water is more calm and the tadpoles are finishing their development. These brooks harbour many different tadpoles, which belong either to *Boophis* or *Mantidactylus*. Up to 15 species could be collected from one brook. Since fish are rare in these upland streams, the tadpoles have apparently a good chance of survival. Adaptation to various microhabitats seems to explain a great deal of the coexistence of many related forms.

Most characteristic for fast running water are the torrent tadpoles of *Boophis* species, such as *B. erythrodactylus* (Figs. 5.4, 5.5, 5.6) and *B. majori* which can apply their greatly expanded beaks as a sucker to rocks and stones. They can be scooped up in shallow rapids among *Hydrostachys* vegetation. They have streamlined bodies, and narrow tail fins. More common are tadpoles of the familiar polliwog type, with rather normally developed mouthparts. Their tail is muscular, enabling them to dart about quickly. These are also *Boophis* species (*B. rappiodes*, *B. luteus*, *B. untersteini* etc.) which live in more quiet parts of streams. They feed on detritus, waterplants and algal crusts.

Although superficially resembling the previous group, the tadpoles of Mantidactylus found in streams have very weakly developed beaks and teeth, or no teeth at all. Apparently they feed on debris and the like supplied by the streams.

The tadpole of *Mantidactylus opiparis* (Figs. 5.7, 5.8) has enlarged mouth flaps richly provided with papillae of different size and the horny beak is reduced. The tadpole keeps itself afloat by supporting itself in plants and twigs, mainly by means of its muscular tail and feeds on small particles on the surface of the water, which are collected with the extended flaps of the mouth. The tadpoles of *M. femoralis* (Figs. 5.9, 5.10) with some minute teeth and degenerate beak live along the shores. The tadpoles of *M. aglavei* dwell above the bottom in dense aquatic vegetation, also holding themselves mainly by bending of their muscular tails.

Species developing in the clear forest brooks have a clear preference for breeding there. But in the forest and adjacent lands there also exist slowly-running shallow muddy waters in boggy places, often as branches of brooks and ponds, which are in fact not very distinct from shallow pools. Species which develop in those slowly-running waters are often also found in pools.

The aquatic species *Mantidactylus betsileanus*, sometimes mixed with *M. biporus*, is very common in these boggy places in or near the forest. Both look like ordinary brown Rana-species, but reach a size of less than 4 cm. One soon gets familiar with their softly sounding, low pitched call, somewhat like a creaking door, which is heard by day and at dusk all the year round. It seems that *M. betsileanus* breeds at any time, the eggmasses of about forty eggs are attached to putrified leaves or other objects on the ground near water. The whole development takes about 2 – 3 months.

Boophis madagascariensis (Fig. 5.11) is another common species associated with shallow smoothly flowing water, but only in the forest. It is a large, up to 8 cm, lean brown treefrog with goggling eyes, large finger discs and small triangular flaps on elbow and heel. It is a frenzied climber. Wild specimens are

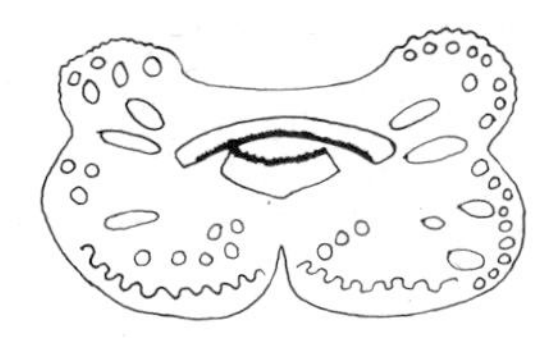
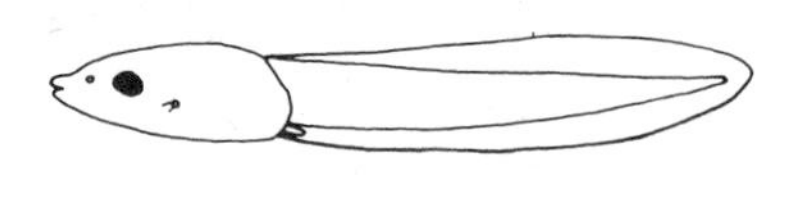

Figs. 5.7, 5.8 *Mantidactylus opiparis*, surface feeding tadpole, mouth and lateral view (L. Blommers)

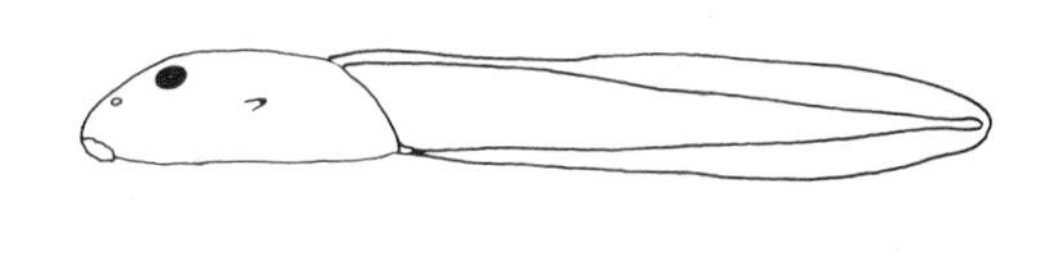

Figs. 5.9, 5.10 *Mantidactylus femoralis*, tadpole running water, mouth and lateral view (L. Blommers)

Fig. 5.11 adult *Boophis madagascariensis*

impossible to keep in a small vivarium, in which they continuously search, but apparently never find a sufficiently high resting place. On rainy summer evenings you hear their groaning calls in the shrubs about one metre above the water. The black eggs are singly laid in water. Two peaks of metamorphosis occur; the largest just before the beginning of the winter, a smaller the following springtime. The young (Fig. 5.12), 2 cm in size, in no way resemble the adults. They are beautifully light green with clear white spots along the head and nice brown reticulations on the back. Being so different they were described formerly as a distinct species. They grow very quickly and within three months they already resemble the adult.

The large frog *Boophis goudoti* (Fig. 5.13), which reaches a size of 10 cm, is not particular in choosing its breeding site; tadpoles are found in pools as well as in running waters. They dwell in the forest but also outside the forest in ricefields and swamps. The adults are often found in water and collected for their legs. Pairs, male and female, continue calling even when in axillary amplexus, and during day and dusk. The call of the male is a low pulsed note and of the female a low buzzing sound. The male repeats its call more frequently than the female.

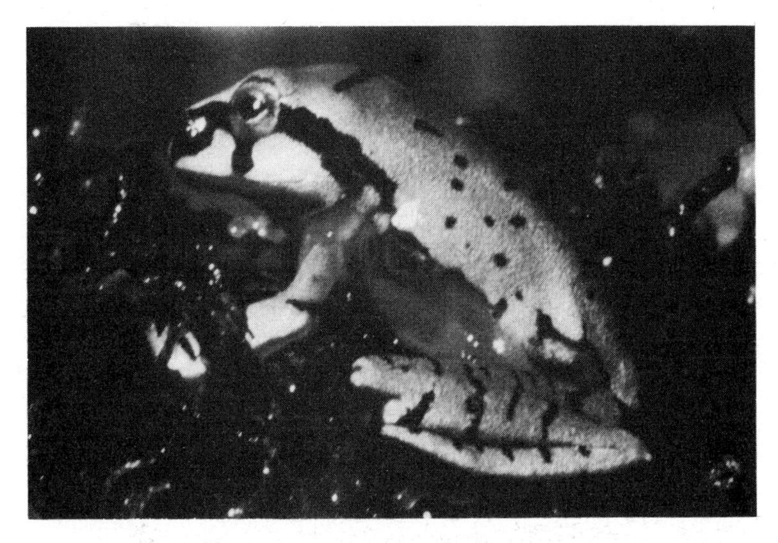

Fig. 5.12 *B. madagascariensis*, young

Fig. 5.13 *B. goudoti*, amplexus (axillary) (L. Blommers)

3.2 Stagnant water

When the first rains refill the ponds and create all sorts of temporary pools and puddles, the forest is suddenly full of frogs. After months of virtual silence, one hears, almost everywhere near stagnant water each cloudy afternoon *Heterixalus* species softly calling and *Mantidactylus blommersae*, too. Most species join them later, at dusk, congregating in choruses of males, which call for most of the night. The days remain almost silent, except for a few places where a rare Mantella- species, brightly coloured and more or less poisonous, may be heard.

Most species breed in large sunlit ponds, in which a luxuriant vegetation of waterplants comes to life. In summer, around New Year, however, tadpoles of some species are even seen in more temporary shallow sunlit puddles, where only some algal bloom may be surmised, and among last season's leaves in acid water in the shade of heavy trees or shrubbery.

Most species mate at night. Couples of *Boophis*- and *Heterixalus*-species are often seen in (axillary) amplexus; the male on top of the female, holding her tightly with his arms. The failure to observe such a prolonged embrace in *Mantidactylus*-species, led us to the observation that the mating contact in species of this genus is very short indeed. When both sexes meet — she attracted by his song — a few seconds of shuffling is sufficient for them to reach the right positions, the male "standing" halfway over the female his thighs touching her head and shoulders (Fig. 5.14). Once in this posture, the female starts oviposition almost immediately, while the male sheds sperm at the same time. The whole act is usually finished within 10 minutes. The male departs often before the last egg is laid, to resume calling a few metres away. Such a furtive mating contact is not known from other frogs except 3 species of the Mantellinae. A feature most likely associated with this peculiar behaviour is the presence of glandular tissue patches at the underside of the thighs of the male, the femoral glands. This and more circumstantial evidence, indicates that a similar mating behaviour is common to the Mantellinae as a whole.

Aglyptodactylus madagascariensis and *Boophis*-species lay their eggs in water, as far as is known in loose batches of at most a few hundred eggs. (In fast running water, the eggbatches of *Boophis* are more consistent and fixed to some support.) The eggs of *Heterixalus*-species are adhered in loose clumps to plants just above the water-surface, and become eventually submerged with further rainfall. Real eye catchers are the eggmasses of some *Mantidactylus*-species on plants overhanging the water. They consist of an initially rather stiff jelly cake containing 30 – 90 eggs. The jelly may be clear or opaque. The green or black eggs of *M. blommersae* are at most some 30 cm above water level, on small bushes and semi-aquatic plants, such as *Crinum* sp., the green eggs of *M. liber* (Figs. 5.15, 5.16), the white eggs of *M. depressiceps* and the green eggs of *M. tornieri* (Fig. 5.17) are usually at 1 m or more on leaves of shrubs and trees. While the eggs develop, the jelly becomes less viscous. The developing tadpoles hang with

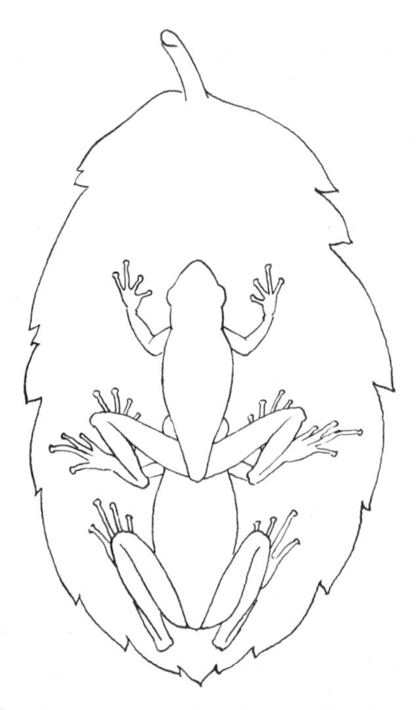

Fig. 5.14 Mating posture in *Mantidactylus liber* (L. Blommers)

Fig. 5.15 *M. liber*, male (L. Blommers)

Fig. 5.16 Eggmass *M. liber*

the tails downwards and finally, after about a week, they drop into the water below, where they develop like ordinary tadpoles. In ground-breeding Mantellinae, such as the *M. ulcerosus* — group of species and *Mantella aurantiaca* eggmasses are similar, and hatching tadpoles reach the water aided by the rain.

In stagnant water, most tadpoles are of the common type, being generalized benthonic feeders with normal beaks and teeth, though their tail is generally less muscular and their fins are higher than in the stream-breeding species. An exceptional beauty is the tadpole of *Microhyla palmata*. The entire body, except eyes and gut is transparent; it floats freely continuously inhaling and filtering. In other words, it is a typical nectonic macrophagous microhylid type of tadpole. Most tadpoles finish development before the end of the summer, not risking complete desiccation of their puddle. A few can wait in permanent ponds for the following spring.

Fig. 5.17 *M. tornieri*, male

3.3 Arboreal nesting sites

The dry winter season is for most frogs a time of silent survival. A notable exception are the species breeding in arboreal sites, or phytotelmes, waterfilled places in trees or in the leaf axils of such plants as Pandanus sp., *Typhonodorum lindleyanum* and *Ravenala madagascariensis*, the Travellers' Palm. Two groups exploit these breeding places: some Cophylinae and *Mantidactylus pulcher* and related species.

Near Perinet, *Platypelis grandis* (Fig. 5.18) presents an example of the arboreal breeding habit. Its far-reaching call can be heard in September and October in afternoon and at night. It resembles the sound of wood cutting, but in a lazy fashion, for the blows are often minutes apart. Being large, the male 9 cm and female 5 cm, this species needs quite large treeholes, or the older somewhat open leaf axils of the Travellers' Palm to breed. If threatened, the male inflates himself and tries to bite, his skin secreting a whitish sticky substance. The female leaves the site a few days after oviposition, never to return. The male cares for the brood of about 100 eggs, staying permanently with it until the tadpoles have changed into froglets. The large eggs contain much yolk, apparently the only source for the development of the free-swimming larvae. It takes some 5 weeks for the development to be completed. The size of the young frogs is only 8 mm. Although the tadpoles lack the ability to feed, rearing them in the absence of the father appears impossible. They die of mould within a few days, suggesting that the father provides not only a bodily but also a chemical protection.

P. grandis is to a great extent representative of the species belonging to the subfamily of the Cophylinae. This group contains treefrogs and burrowing frogs with a size of 10 – 100 mm. The larval development hitherto known is always the same; the larvae are free-swimming and nonfeeding and it is the father who cares for the brood. The treefrogs breed in arboreal nesting sites in the wet and less so in the dry season, while the burrowing species rear their progeny in a small water-filled hole in the ground only during the wet season. Feeding on ants seems widespread in this group. Another species regularly met is *Plethodontohyla notosticta* (Fig. 5.19), which has a preference for breeding in the Travellers' Palm.

Fig. 5.18 *Platypelis grandis*, male and female (L. Blommers).

Anodonthyla boulengeri, a tiny frog, fits perfectly in the narrow tube formed by the two folded edges on the leaf stalk of the Travellers' Palm and finds there a hiding place.

Within the genus Mantidactylus there is a group of small brightly green or yellow coloured treefrogs, the *M. pulcher* group, which live and develop in the so-called aerial swamp. You find them mostly at the base of prickly Pandanus leaves, even in the dry season together with their tadpoles, which are well adapted to survive in these small waterplaces. The body is very flattened and the tail very muscular, with the fin reduced. The horny beak is broad and robust and they have many teeth, which they use for scraping the algae from the leaves (Figs. 5.20, 5.21). The tadpoles wriggle about in the water at the leaf bases and they are capable of leaving the water by violent wiggling of the tail and may move that way to the next lower axil, when the water in their axil is about to disappear.

Fig. 5.19 *Plethodontohyla notosticta*, eggs, tadpoles and young (L. Blommers)

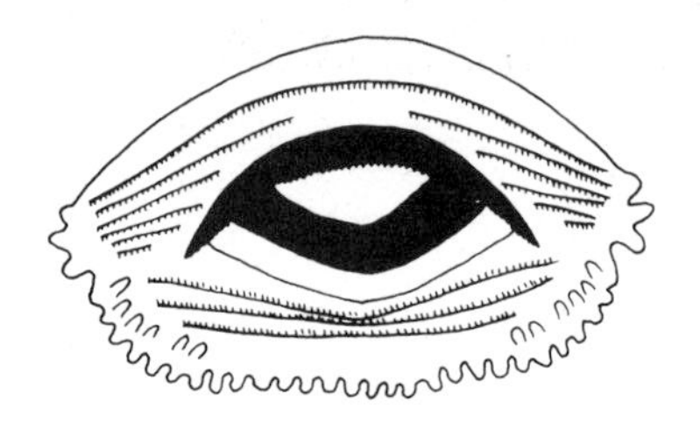

Figs. 5.20, 5.21 *M. pulcher* group, tadpole in axils of Pandanus, mouth and lateral view (L. Blommers)

3.4 Development outside the water

Mantidactylus asper and *M. eiselti* are heard calling far away from water in the rainy season. *M. asper* calls on treetrunks and in shrubs in the late afternoon by cloudy weather. The calling males are tens of metres removed from each other and from water. It was impossible to find their tadpoles. Once a single egg was discovered attached to a dead mossy branch in a shrub about 1 metre above the ground, near the calling site of a male. After 14 days, an embryo with black eyes, a narrow body with four stumps and a tail had developed inside the egg membrane. Thus, there remains little doubt about the direct development outside the water in *M. asper*. *M. eiselti* is found on hills in the forest, sometimes hundreds of metres removed from water. It calls during the day, sitting in dense shrubbery just above the ground.

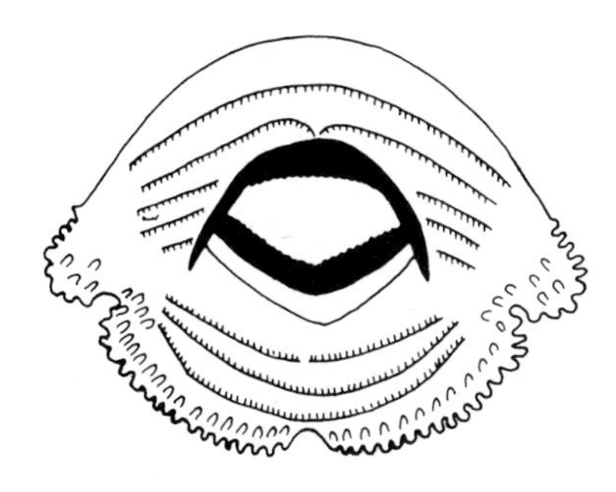

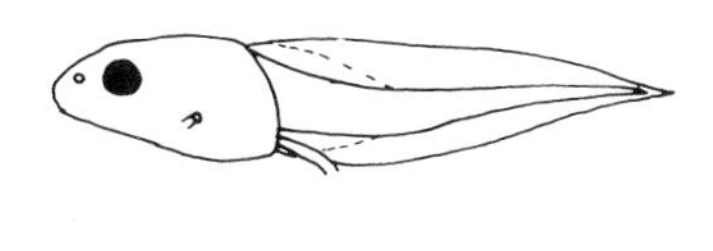

Figs. 5.22, 5.23 *Boophis tephreaomystax*, tadpole of stagnant water, mouth and lateral view (L. Blommers)

Fig. 5.24 *Psuedohemisus granulosus*, female (L. Blommers)

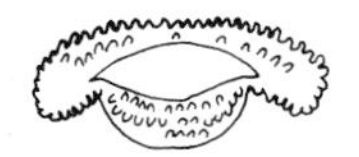

Fig. 5.25 *Pseudohemisus,* mouth of tadpole (L. Blommers)

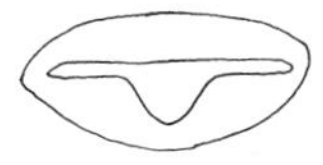

Fig. 5.26 *Dyscophus,* mouth of tadpole

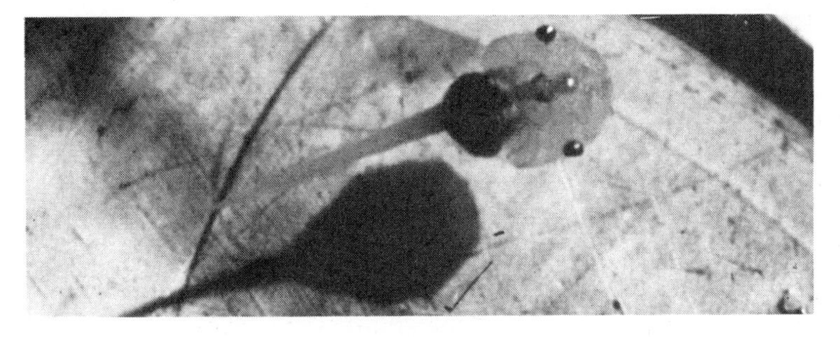

Fig. 5.27 *Dyscophus quinquelineatus*, tadpole (L. Blommers)

The calling males are far removed from each other, and very difficult to catch, since they stop singing immediately when approached. Although we did not find the eggs, it seems that larval development in this species also occurs on land.

4. THE WESTERN DECIDUOUS FOREST

The western part of Madagascar receives less precipitation than the opposite side, and, probably even more important to frogs, rains fall usually during the entire winter season, up to 8 months in one stretch. Therefore, it is not very amazing that 5 species, all burrowing and belonging to the genera *Dyscophus, Pseudohemisus* and *Tomopterna*, are exclusively known from this part of the island. All other species found there have a distribution which includes at least part of the East. It should be realized however, that the West is also much less explored, since the area is hardly accessible during the rainy season.

An exception forms the Ankarafantsika forest near Ampijoroa on the route to Majunga. It is a beautiful deciduous lowland forest, which is crisp dry at the end of the dry season and as it seems completely devoid of amphibian life. This impression is belied as soon as the first thunderstorms arrive. Evoked by the slashing rains, frogs turn up everywhere, notably at night when burrowing species move around in numbers. From the trees a sound like the barking of a small dog indicates that *Boophis tephraeomystax* is present almost everywhere near water, while the lower vegetation is full of the calls of *Mantidactylus wittei* and *Heterixalus betsileo*. As the rain gives rise to numerous ponds and puddles, breeding soon starts and by January nearly every larger pool is full of tadpoles of *M. wittei*, *B. tephraeomystax* (Figs. 5.22, 5.23), *H. betsileo* and *Ptychadena mascareniensis*.

, *Dyscophus quinquelineatus* and *Pseudohemisus granulosus* (Fig. 5.24) breed often in rain puddles. Their microphagous tadpoles develop fast, *Pseudohemisis* (Fig. 5.25) in about 3 weeks and *Dyscophus* in 2 months, and even so, they run the risk that their shallow puddle may dry up too soon. The tadpole of *D. quinquelineatus* (Figs. 5.26, 5.27), a typical microhylid, floats horizontally in the water, motionless except for a slight undulation of its flimsy tail and steady snapping of the mouth. Most interesting is the larva of *P. granulosus*. It is intermediate between the typical microhylid tadpole, as in *Dyscophus* and the familiar polliwog. Just like the adult of this and other species of the Scaphiophryninae, it combines features of Microhylidae and Ranidae. It is a skilful swimmer, which easily turns upside down to feed on particles on the surface, or searches the bottom almost like a fish with the tail obliquely upwards. Although without teeth or horny beak, it also feeds on dead congeners.

5. WHERE THE FOREST HAS GONE

Few Malagasy frogs are found in open country. *Ptychadena mascareniensis* — the only Afro-Malagasy species — imported *Rana tigrina*, and *Tomopterna labrosa*, endemic but with close relatives in the African savannah, are the first to come to mind. Being robust and agile ground dwellers, the first two are well adapted to a man-made landscape, with ricefields and other plantations. *T. labrosa* lives as burrowing species even in parts of the southern region. It is often found in and near washing ponds and irrigation channels in the lowland around Tulear. Of the treefrogs, *Heterixalus* species (Fig. 5.28) and somewhat less, *Boophis tephraeomystax* are best adapted to sunshine, in which they get a much lighter complexion. A few aquatic species, notably *Mantidactylus betsileanus*, *M. alutus* and *Boophis goudoti* seem able to survive in and around wet ricefields and the like. They are found there regularly notwithstanding probably strong competition by *P. mascareniensis*. Finally, some burrowing microhylids may appear quite resistant to drought. In other words, only some 10% of the known 150 species thrive or merely stay alive where man dominates nature. Another 45 species or 30% live in one of the three high mountain areas. For

Fig. 5.28 *Heterixalus tricolor*, male (L. Blommers)

the remaining majority of Malagasy frogs the forests, at low or medium altitude, are the only suitable environment. It is in these forests that most of the vast adaptive radiation has occurred. Some species, like *Mantella aurantiaca* and *Dyscophus antongilii* which are well known by amateurs and sometimes displayed in a zoo, seem almost restricted to a single locality. In such cases both the local preservation of species and its habitat, and the already existing rules prohibiting international trade are needed to prevent extinction.

Many species are probably rather more widespread. Unfortunately, we are still almost ignorant about the actual geographic distribution of many, but the occurrence of a great part (30%) of known species in such limited areas as Perinet or Mandraka valley suggests for example that most of these species are rather widely distributed in the forest on the eastern escarpment. But while this is one of the best preserved areas, the eastern coastal rainforest, the western deciduous forest and the Gallery forests on the central Highland (Tampoketsy) are hardly more than last remnants of biotopes on which some species depend. Even so, the forests and brooks of each mountain area constitute the entire living space for some unique species. Therefore, conservation of forests in every region is essential to preserve the unique variety of Malagasy anurans.

REFERENCES

Arnoult, J. (1966) Ecologie et developpement de Mantella aurantiaca. *Bull. Mus. natn. Hist. nat. Paris* 37 (2), 931 – 40, figs.

Arnoult, J. and M. Razarihelisoa (1966) Contribution a l'etude des Batraciens de Madagascar. Ecologie et forme larvaire de *Rhacophorus* goudoti (Tschudi). *Bull. Mus. natn. Hist. nat. Paris.* 38 (2), 112 – 27, figs.

Arnoult, J. and M. Razarihelisoa (1967) Contribution a l'etude des Batraciens de Madagascar du genre Mantidactylus: adultes et formes larvaires de M. betsileanus (Blgr.), M. curtus (Blgr.) et M. alutus (Peracca). *Bull. Mus. natn. Hist. nat. Paris.* 39 (2), 471 – 87, figs.

Blommers-Schlösser, R.M.A. (1975a) A unique case of mating behaviour in a Malagasy treefrog, Gephyromantis liber (Peracca, 1893), with observations on the larval development (Amphibia, Ranidae). Beaufortia 23, 15 – 25, figs.

Blommers-Schlösser, R.M.A. (1975b) Observations on the larval development of some Malagasy frogs, with notes on their ecology and biology (Anura: Dyscophinae, Scaphiophryninae and Cophylinae). Beaufortia 24, 7 – 26, figs.

Blommers-Schlösser, R.M.A. (1979a) Biosystematics of the Malagasy frogs. I. Mantellinae (Ranidae). Beaufortia 29, 1 – 77, figs.

Blommers-Schlösser, R.M.A. (1979b) Biosystematics of the Malagasy frogs. II. The genus Boophis (Rhacophoridae). *Bijdr. Dierk.* 49 (2), 261 – 312, pls.I-IV.

Blommers-Schlösser, R.M.A. (1981) On endemic Malagasy frogs (Ranidae, Rhacophoridae and Hyperoliidae). *Mon. zool. ital.* 15 (12), 217 – 24.

Guibe, J. (1978) Les Batraciens de Madagascar. *Boon. Zool. Monogr.* 11, 1 – 140, pls. I-LXXXII.

Guibe, J. (1952) Recherches sur les Batraciens de Madagascar. *Mem. Inst. Sci. Madagascar* 7 (1,A), 109 – 16.

Milot, J. and Guibe, J. (1951) Batraciens malgaches a biotope vegetal. *Mem. Inst. Sci. Madagascar* 5 (1,A), 197 – 212.

Razarihelisoa, M. (1973) Contribution a l'etude des Batraciens de Madagascar: Ecologie et Comportement fouisseur du "Sahondoroka" Plethodontohyla tuberata Peters, batracien Anoure endemique du Centre de Madagascar. *Ann. Univ. de Madagascar* 10, 103 – 25.

Razarihelisoa, M. (1974a) Contribution a l'etude des Batraciens de Madagascar: Ecologie et developpement de Gephyromantis metheuni Angel, batracien a biotope vegetal sur les Pandanus. *Bull. Acad. Malg.* 51 (1), 113 – 28.

Razarihelisoa, M. (1974b) Contribution a l'etude des Batraciens de Madagascar: Ecologie et developpement de Mantidactylus brevipalmatus Ahl, batraciens des eaux courantes. *Bull. Acad. Malg.* 51 (1), 129 – 42.

Razarihelisoa, M. (1979) Contribution a l'etude biologique de quelques Batraciens de Madagascar. *These Doctorat d'Etat Paris* VII, 1 – 195.

TABLE I. Different life strategies known in Malagasy frogs.

Eggs	Tadpoles	Frogs	Names	References
In water	In stagnant water	Arboreal	*Boophis tephraeomystax* group*	Bl-Schl., 1979b
In water	In stagnant water	Arboreal (aquatic)	*Boophis goudoti*	Arnoult & Raz., 1967; Bl-Schl., 1979b
In water	In stagnant water	Arboreal	*Boophis madagascariensis*	Bl-Schl., 1979b
In water	In stagnant water	Terrestrial (aquatic)	*Rana tigrina, Ptychadena mascareniensis*	Raz., 1979
In water	In stagnant water	Terrestrial (aquatic)	*Aglyptodactylus madagascariensis*	Bl-Schl., 1979a
In water	In stagnant water	Burrowing	*Tomopterna, Dyscophus, Pseudohemisus*	Bl-Schl., 1975b
In water, collected in plants or trees	Freeswimming, nonfeeding larvae; the father with the direct developing tadpoles	Arboreal	*Plethodontohyla notosticta, Anodonthyla boulengeri, Platypelis grandis, P. tuberifera, P. pollicaris, Paracophyla tuberculata*	Millot & Guibe, 1951; Bl-Schl., 1975b; Raz., 1979
In a water-filled hole in the ground	Freeswimming, nonfeeding larvae; the father with the direct developing tadpoles	Burrowing	*Plethodontohyla tuberata*	Guibé, 1952; Raz., 1973
In water	In running water	Arboreal	*Boophis rappiodes* group, *B. luteus* group, *B. rhodoscelis* group, *B. untersteini*	Bl-Schl., 1979b
In water	In running water	Terrestrial (aquatic)	*Boophis microtympanum, B. williamsi*	Bl-Schl., 1979b
Just above water	In stagnant water	Arboreal	*Heterixalus*	Raz., 1979; Bl-Schl., 1981
Just above water	In water, collected in leaf axils	Arboreal	*Mantidactylus pulcher* group	Bl-Schl., 1979a; Millot & Guibé, 1951; Raz., 1974a
Outside water, on ground	In stagnant water	Terrestrial	*Mantella*	Arnoult, 1966; Raz., 1979
Outside water, on ground	In stagnant water	Terrestrial (aquatic)	*Mantidactylus alutus, M. ulcerosus*	Bl-Schl., 1979a
Outside water, on leaves	In stagnant water	Arboreal	*Mantidactylus liber, M. depressiceps* group, *M. wittei* group	Bl-Schl., 1979a
Outside water, on ground	In running water	Terrestrial (aquatic)	*Mantidactylus lugubris* group, *M. opiparis* group, *M. curtus, M. betsileanus*	Bl-Schl., 1979a; Raz., 1974b
Not known	In running water	Arboreal	*Mantidactylus aglavei*	Bl-Schl., 1979a
Outside water	Direct development within eggmembrane	Arboreal	*Mantidactylus asper, M. eiselti*?	Bl-Schl., 1979a

* species groups as defined by Blommers-Schlösser, 1979a + b;
† Bl-Schl.: Blommers-Schlösser; Raz.: Razarihelisoa

CHAPTER 6

The Reptiles

CHARLES P. BLANC

Malagasy reptiles and amphibians have been the subject of particularly active recent research: taxonomic, biological and biogeographical. Many new species have been described: 30 amphibians between 1973 and 1976 compared to only 7 found in the preceding quarter century, and as many reptiles, mostly from my own prospecting in various regions of the Island. This chiefly proves that the list is still far from complete.

Nonetheless, even in our present state of knowledge, the 260 species of reptiles are a very large number compared to the modest size of the country, only 587,000 km^2. Furthermore, 95 – 99% of them are endemic to Madagascar or the Malagasy region. They are unequally divided among 3 orders: one species of crocodile, 13 species of Chelonia or tortoises, and about 240 species of Squamata, comprising 180 lizards and 60 snakes.

This unique fauna, which has very ancient origins, reflects the stages of the history of life in Madagascar — the evolutionary steps which have turned this country into the famous "naturalists' promised land". The reptiles like various other Malagasy radiations show:

— relative species richness on a basis of taxonomic poverty, each major group being represented by few families or genera,

— sharp discordance between the faunal composition and that of neighbouring continents and islands,

— a high percentage of endemism at species level but no endemic families,

— contrasting patterns of speciation. On the one hand we find species-poor genera, mainly endemics, relicts of ancient evolutionary lines for whom Madagascar has been an island refuge. On the other we find highly diversified genera which may contain more than 30 species, whether endemic or not. For these genera Madagascar has acted as a mini-continent.

The reptiles have predominantly African affinities. Fossils are little use in pinpointing their origin. Some are too old — Permian or lower Triassic, — while others are extinct giant forms — Cretaceous dinosaurs and Boidae.

These characteristics of the herpetofauna (amphibians and reptiles) imply early and permanent isolation. Most of the fauna results from chance immigrations on floating rafts of vegetation, at different times. They also imply that there was feeble competition and low selection pressure, at least for a time, after the invasions. Madagascar thus serves both as a vast natural laboratory particularly suited to research on speciation, and as a sanctuary for species which have disappeared elsewhere tens of millions of years ago.

These species are of great scientific interest and are nearly always insufficiently known. They are also very vulnerable to human action on the habitat. The situation is even more worrying because many of them have point distributions, like *Geochelone yniphora*, the plowshare tortoise, at Cape Sada, or the gekkonid lizards *Lygodactylus arnoulti* and *L. blanci* which both live only on the summit of Mount Bity. The local situation may grow even worse with the disappearance of the "fady", taboos which used to protect large tortoises and boid snakes. It can also be aggravated by commercialization and local hunting for food, particularly for crocodiles and freshwater turtles.

An inventory of urgent conservation needs includes species censuses (both numbers and distributions) and protection measures such as those proposed by the WWF for *Geochelone yniphora*. Every year lost could close with the extinction of species, some of which will never have been identified. We cannot insist too much on the importance of discovering all the biological components affected by protection measures. Just to quote one example, permission to collect dead wood from the forest has a large negative impact on the burrowing frog *Plethodontohyla tuberata* which hollows out its egglaying chambers in large decomposing branches and trunks.

CROCODILES

Naturalists may deplore the near eradication of *Crocodilus niloticus*, the "voay" or "mamba", from the lakes and rivers of Madagascar. It has been exterminated for commercial sale of its skins, with the active cooperation of villagers who are only too glad to be rid of this malign and dangerous neighbour. Egg-collecting and destruction of the young have added to its demise. The increase in human security has, however, upset the previous ecological equilibrium.

The last specimens are now confined to a few regions where access is difficult, such as the subterranean rivers of the Ankarana, or where crocodiles are protected by a taboo, "fady", in a few sacred lakes, like Lake Anivorano between Diego Suarez and Ambilobe. There cattle may even be sacrificed to the crocodiles during certain festivals.

TORTOISES AND TURTLES

The most famous tortoises are the 5 terrestrial species of the family Testudinidae. Only one is not endemic, *Kinixys belliana* which was probably imported from eastern or southern Africa. It is now common in the northwest, near Ambanja and on Nosy Faly island. The four others are limited to the southern and western domains. Each domain has one large and one small form.

In the south, the radiated tortoise, *Geochelone radiata* or "sokaka" is the best known. It has a domed, nearly spherical back with bright yellow stripes that radiate in star pattern from the centre of each dorsal plate, on a black background. The largest individuals reach 60 cm in length. It lives sympatrically with *Pyxis arachnoides*, the "kapila", which is much smaller. The kapila's plastron has a hinged shutter in front which closes to protect the head like an American box turtle's.

In the west there lives another species of the genus *Geochelone*: *G. yniphora* or "angonoka", the plowshare tortoise. It is even larger than *G. radiata*, with a pronounced gular spur from the front of its plastron. It is very near final extinction: only a few individuals survive near Soalala. *Acinixys planicauda*, the "kapidolo" lives in the Morondava region. This is a moderate-sized tortoise easy to identify by its carapace in three distinct sections: the scales of the middle row are flat and horizontal, with steeper sloping sections either side.

There are three freshwater turtles of the family Pelomedusidae. *Erymnochelys madagascariensis*, the "rere", has left fossils in Africa from the Upper Cretaceous. It is closely related to a South American genus *Podocnemis*, with 7 species. Its carapace length can exceed 50 cm. This relict species has survived in Madagascar, but its refuge is more and more threatened as the edges of lakes and large rivers of the

Fig. 6.1 Nile crocodile, Berenty (A.M. Jolly)

west and north – central regions are converted to ricefields, which reduces its sites for egg-laying. The two other, smaller species *Pelusios subniger* and *Pelomedusa subrufa* (''kapika'') have a range which includes Africa. The first is identifiable by its hinged anterior plastron. They are both often boiled and eaten.

The sea-turtles on the Malagasy coasts are:

— The green turtle, *Chelonia mydas*, ''fanojato'', which is much hunted for its meat. Decary described the Sakalava ritual: the harpooned animal is carried on a litter of branches and then dismembered alive. Its head is impaled on a stake above the heads of previous victims.

— The hawksbill, *Eretmochelys imbricata* or ''fano hana'', which is hunted for tortoiseshell. Its populations are lower than before, and, above all, the size of captured animals has greatly decreased.

— The luth-turtle *Dermochelys coriacea*, ''valozoro'', is occasional.

— Ridley's turtle *Lepidochelys coriacea* is also occasional or rare.

Fig. 6.2 *Geochelone radiata* (Ph. Oberlé)

LIZARDS AND CHAMELEONS

The 180 lizard species belong to only 5 families: Gekkonidae, Iguanidae, Gerrhosauridae, Chameleontidae and Scindidae. Other families which are well represented in Africa, Asia and Australia have not reached Madagascar: the Agamidae, Varanidae, Lacertidae, etc. The Malagasy Iguanidae are surprising: they are essentially an American family which have no living or fossil representatives on the continents surrounding Madagascar.

Among the Gekkonidae, let us begin with the genus *Uroplatus*. This is an endemic group, highly specialized, and characterized by its extraordinary mimicry of trunks and lianas. The animal clings immobile, head downward: its skin perfectly imitates the irregularities of bark and even lichens. A skinfold with tattered edges runs along its side and limbs from jaw to tail, and merges with the treetrunk, casting no shadow. Only the most meticulous observation will lead to its discovery. The fingers enlarge into discs at the end. One of the best-known species, *U. fimbriatus*, the ''taha-fisaka'', lives in the eastern forest and can reach 29 cm total length. Flacourt claimed in 1658 that the uroplates' fingers adhered so strongly to human skin that you could only remove the animal by cutting off the piece of skin it has seized with a razor blade! That is legend, of course, but the fingers do act as remarkable suction cups.

The genus *Phelsuma* includes magnificent geckos. Their basic colour is brilliant emerald, picked out with red dots or longitudinal yellow or white lines. *Phelsuma* has undergone intense speciation in Madagascar. Its distribution extends to neighbouring islands: the Comoros, Seychelles, Mauritius, Reunion, and even the Andaman Islands. *P. madagascariensis* is a large form, variable in colour, and

Fig. 6.3 *Phelsuma madagascariensis*, Nosy Bé (A. Jolly)

common in the east and north; *P. lineata* is 10 – 12 cm total length, abundant in banana plantations, *P. barbouri* is adapted to rocky habitats.

Lygodactylus, small lizards 6 – 8 cm long, greyish spattered with black, have peopled Madagascar from Africa and are differentiated into 3 groups linked to each of the great phytogeographic divisions: west, south and east.

We may also note the endemic genus *Geckolepis* whose scales are flat, smooth, rounded and overlapping like the scales of a fish — and are also very fragile.

The Iguanidae — whose presence in Madagascar is a biogeographic mystery — have two endemic genera limited to the south and west of the country. The 6 *Oplurus*, "sitry" or "androngo", are fairly large, 20 – 30 cm, and are part of the Malagasy landscape. You cannot miss seeing them running agilely on rocks and treetrunks. The other genus has just one species, *Chalarodon madagascariensis* ("langalia"). It is very common on sandy soils, pretty and tame, with a well developed third eye haloed in black on the top of its head. During the mating season the more aggressive males put on bright nuptial coloration and defend territories. Frontiers are decided by ritualized threat movements, which sometimes degenerate

Fig. 6.4 Malagasy chameleon (A. Jolly)

Fig. 6.5 Malagasy chameleon (A. Jolly)

into real combat. The two opponents grab each other by their jaws, and turn double somersaults without letting go.

The Gerrhosauridae also have two endemic genera which differ in their speciation. *Tracheloptychus* has 2 closely related species, *Zonosaurus* 10 species. The former is confined to the south and southwest, the latter has peopled all geographic regions. One, *Z. madagascariensis* has even implanted itself on the Glorieuses and Aldabra. The giant species, *Z. maximus*, about 60 cm long, has aquatic habits.

For the two genera of Chameleontidae, *Chamaeleo* ("tanala, sangorita, tarondro"), and *Brookesia* ("anjava"), recent discoveries lead us to recognize 32 and 19 species respectively. Their centre of origin seems to be East Africa, but Madagascar contains two thirds of the world's species.

Chameleons are highly specialized for life on branches and tall plants, thanks to the transformation of their hands and feet into pincers with two toes opposing three others. They are equally specialized for stalking prey: their globular eyes in conical eyelids, which pivot independently, can explore nearly all the surrounding space while the head stays immobile. Their tongue can strike accurately and instantaneously at an insect nearly a body's length away, and capture the insect on the swollen, slightly flattened, sticky tongue-tip. Their characteristic hesitating walk allows them to judge distances, and has given rise to several Malagasy proverbs: "The chameleon counts his steps wherever he goes." When they are in danger they rapidly disappear among the branches, hanging by the prehensile tail.

Madagascar is home for the largest species of the genus: *C. oustaleti*, 68 cm long, and the smallest: *C. nasutus*, 10 cm. Chameleon species differ in the presence of nasal horns, of crests and helmets on the top of the head, and of occipital lobes which cover the neck in a frill or cape. This armour relates to the great intolerance between individuals. They are also famous for their beauty of colouring and its rapid changes, which correspond with their degree of aggressivity.

The commonest species near Tananarive, *C. lateralis*, owes its abundance to its great fecundity. The Eastern forests have gorgeously coloured species: *C. parsoni* and *C. pardalis*.

The *Brookesia* are much smaller. They carry bony spines on the back, and live in forests. Some, generally blackish in colour, live on the ground in dead leaf litter (*B. nasus*). Others, grey or green, live on mossy trees.

Of the 48 described species of Scindidae, only one is not endemic: *Ablepharus boutonii*, a little lively black lizard of rocky beaches. Even this has two endemic subspecies, one on Madagascar itself, one on Nosy Bé. The Scindidae are divided unevenly among 10 genera. 2 with pantropical distributions, 2 with mainly African distribution (*Acontias* and *Scelotes*), and 6 endemic. These last are monotypic or with very few species. They have burrowing habits, which gives their local name, "Matahotr' andro". They form a series of morphological steps in the degeneration of limbs down to the limbs' complete disappearance, particularly in the genus *Grandidierina*.

The genus *Scelotes* may be a relatively recent arrival, but it has diversified into 25 species. They are all endemic, some very ill-defined. The more remarkable ones are an aquatic form, *S. astrolabi*, the largest, (50 cm long), *S. splendidus* whose back has a succession of transverse black bars on a light yellow background, and *S. igneocaudatus*, fairly small (12 – 15 cm) named for its blazing red tail, whose distribution marks out an ancient peneplain.

The commonest species is *Mabuya gravenhorsti* which has been favoured by the increase in secondary "savoka" vegetation.

SERPENTS

Madagascar has only 3 families of snakes: the Typhlopidae, the Boidae, and the Colubridae. The Elapidae, such as cobras and mambas, and the Viperidae, such as vipers and rattlesnakes, do not exist there. Malagasy snakes are thus harmless to man, which is very reassuring to naturalists who can wander

Fig. 6.6 *Acrantophis dumerilii*, the western boa (A. Jolly)

Fig. 6.7 *Langaha nasuta*, female, with leaf-like prolongation of her muzzle (A. Jolly)

through Malagasy forest without fear.

The Typhlopidae are represented only by the genus *Typhlops*, with 9 wormlike, burrowing, well concealed species. The least rare is *T. braminus*, which is also the only non-endemic one. It lives in humus and piles of rotting vegetation.

The Boidae retain vestiges of their hindlimbs in the form of two spurs on either side of the cloaca, which are used during courtship. They are one of the remarkable components of the Malagasy fauna with their two endemic genera: *Sanzinia* and *Acrantophis*. The single *Sanzinia* species, *S. madagascariensis*, is a tree-boa, small and aggressive. You commonly find it with its body curved in symmetrical rings one either side of its head, living on a tree-branch where its greenish tint and white markings camouflage it. It is easily recognizeable by its high, regularly spaced, supra-labial scales which are separated by a deep furrow. The two species of *Acrantophis*, the "do", are placid, terrestrial snakes. They are 150 – 180 cm long, brownish picked out with black lozenges. They frequent the humid regions near streams and ponds. They are much feared and implacably killed, and their habitat is rapidly disappearing. They are even more vulnerable because of their large size. There should be an attempt to breed them in captivity, which would be fairly easy because they are ovo-viviparous and give birth in captivity without difficulty.

Of the 16 genera of Colubridae, with about 50 species recognized, only one genus (*Geodipsas*) extends to Africa. All the colubrid species are endemic either to Madagascar itself or to the Malagasy geographic area. They are assigned to two groups by their tooth structure: Aglyphes and Opisthoglyphes. The monospecific genus *Mimophis mahafalensis* differs from all other Malagasy Colubrids, and resembles instead the African genus *Psammophis*, by the absence of hypapophyses on the posterior vertebrae and by its thread-like, non-bifurcated hemipenis. It is thin and agile and lives in the Southern bush, where several very pale specimens have been seen.

The distribution and biology of many forms is still little known. Many of them are very discrete or ultra-rare, and several have only been found in a single site.

Among the most remarkable we may cite:

— Genus *Langaha*, famous for its sexual dimorphism. In *L. alluaudi* the female alone carries a jutting sub-ocular scale. In *L. nasuta* the female has a scaly, toothed, flexible leaf-shaped organ flattened transversely but long sagitally, which protrudes from the front of the muzzle. In the male the organ is conical and the upper lip is yellow.

— *Lioheterodon madagascariensis*, the "menarana", which reaches 150 cm long. Its markings are alternate yellow and black, and it terrifies local people. A related species, smaller and lighter-coloured, *L. modestus*, hunts lizards in the south, like *Dromycodryas quadrilineatus* which is easy to identify by its longitudinal stripes.

— *Liophidium rhodogaster* is a forest species, identifiable by its red abdomen and the red lower surface of its tail.

— *Madagascarophis colubrina*, the "renivitsika", about 100 cm long, with thick yellowish body and brown spots arranged quincunx pattern — four spots at the corners of a rectangle with one in the centre. It hides in termite hills and even ant nests.

— *Ithycyphus miniatus*, the "fandrefiala" is the centre of many legends. It is supposed to transfix zebus by dropping from trees, stiff as an arrow, tail first. Its extremely thin metre-long body, its arboreal habits, and its bright red tail presumably gave rise to this bizarre belief.

BIBLIOGRAPHY

Angel, F. 1942 Les Lezards de Madagascar, *Mem. Acad. Malg.*, 36, 194pp.
Blanc, C.P. 1977 Reptiles Sauriens Iguanidae, *Faune de Madagascar* 45, 197pp.
Brygoo, E.R. 1971 Reptiles Sauriens Chamaeleonidae, le genre Chamaeleo, *Faune de Madagascar* 33, 318pp.

Brygoo, E.R. 1978 Reptiles Sauriens Chamaeleonidae, genre Brookesia, et complement pour le genre Chamaeleo, *Faune de Madagascar* 47, 173pp.
Guibe, J. 1958 Les Serpents de Madagascar, *Mem. Inst. Scient. Mad.*, ser. A, 12, 189–260.
Perrier de la Bathie, H. 1914 Les Crocodiles malgaches, *Bull. Acad. Malg.*, t.I, 129–35.

CHAPTER 7

The Birds of Madagascar

CONSTANTINE W. BENSON

INTRODUCTION

Three points must be kept continually in mind: (1) The bird fauna of Madagascar has a relative poverty in numbers of species but remarkable uniqueness, and many striking differences from the birds of Africa, separated by the Mozambique Channel, at its narrowest only about 400 km wide; (2) It follows from (1), the extreme importance of adequate protection, above all by habitat conservation; (3) The contrast in habitats, broadly:

(a) dense evergreen forest in the humid east, and extreme north (Mt. d'Ambre) and northwest (Sambirano), rainfall up to 750 cm per annum,

(b) open savanna in the west generally, rainfall only about 100 cm,

(c) subdesert scrub in the extreme southwest, rainfall less than 40 cm. Through the centre there runs a highland ridge, rising in places to well over 2000 m, but unlike eastern Africa holding almost no special montane birds. The affinity of this area is with the humid east, although its original forest has been replaced largely by cultivation.

It seems that some 70 million years ago (about the end of the Cretaceous) Madagascar was still joined to Africa. But the great majority of the birds would have arrived subsequently, by flying, although this does not apply to the giant, flightless elephant birds (Aepyornithidae), related to the African ostriches (Struthionidae). The largest of these, larger than any ostrich (height more than 3 m, weight 450 kg, size of egg as much as 33 x 23 cm), may have survived in the extreme south until about 200 years ago, but have been extirpated by hunting. Subsequently, the only species which may have become extinct is Delalande's Coua *Coua delalandei*. It inhabited the northern half of the humid east, but there is no record of it since 1834. Although not comparable in size to the Aepyornithidae, being only 57 cm long, it was sufficiently large to have been worth eating.

By no means all the birds have an African origin, for there is a distinct Asiatic element. Access from Indomalaysia would have been much easier 25,000 years ago than now. At that time, in a glacial epoch, ocean levels were some 150 m lower, so that the area of exposed land in the intervening area was much greater. Thus the Seychelles formed a continuous land mass, instead of the fragmented archipelago of today.

THE ENDEMIC FAMILIES

There are 5 families virtually confined to Madagascar. The mesites (Mesitornithidae) are of uncertain affinities, but usually placed in the same order as that of the rails (Rallidae). There are 3 species, the Brown Mesite *Mesites unicolor* in the humid east (particularly difficult to see — one might spend weeks searching for it, without success), one in a restricted area in the north of the western savanna, the White-breasted Mesite *M. variegata* (deserving special protection), and one in the subdesert scrub, the Monias (or Bensch's Rail) *Monias benschi* (the easiest to observe). Each is about 30 cm long, and very terrestrial, although the nest is placed above the ground. The last named lives in groups of up to 20 individuals and is believed to be polygamous, two females sharing the same nest.

Although Milon *et al.* place the next two families with the true rollers (Coraciidae), this is surely too embracing. The courols (Leptosomidae) are represented by a single species, which has also colonized all the Comoro Islands, but regarded as *de facto* endemic. It is easily observed, and relatively common, often first recognized by its striking and beautiful call, made from a conspicuous perch or even in circling flight. It appears to represent a very early colonization by true rollers from Africa (or Asia), perhaps millions of years ago. Later, there was another colonization of roller stock, which evolved into another family, originally entirely endemic. There are 5 species of ground rollers (Brachypteraciidae), 4 of them confined to the humid east, very difficult to observe, entirely dependent on a forest environment. The Short-legged Ground Roller *Brachypteracias leptosomus*, and Scaled Ground Roller *B. squamigera* are not so strictly terrestrial as the Pitta-like Ground Roller *Atelornis pittoides* or Crossley's Ground Roller *A. crossleyi*, which in contrast to the true rollers and courols are almost entirely so. The Pitta-like Ground Roller nests at the end of a 50 cm long horizontal tunnel into the forest floor. Almost certainly this applies to the other three. The Long-tailed Ground Roller *A. chimaera* is confined to subdesert scrub, relatively easy to observe. It too nests in a tunnel into the ground.

Fig. 7.1 *Monias benschi*, the Monias (P. Randriamanantsoa, after A. Grandidier)

Fig. 7.2 *Atelornis pittoides*, the long-tailed ground roller (Rabarijaona, after A. Grandidier)

Fig. 7.3 *Philepitta castanea*, the velvet asity (P. Randriamanantsoa, after A. Grandidier)

The asities and sunbird asities (Philepittidae) compose a family within a large group of primitive song birds (suboscines), only otherwise represented in the Old World by the pittas (Pittidae), of which there are many species in southeast Asia and two in Africa. By far the largest representation of this group, however, is in South America. The Philepittidae consist of two superficially very different pairs of species, although they seem to comprise a single family. The Velvet Asity *Philepitta castanea* is confined to the humid east, and is replaced in the northwest (in the Sambirano) by Schlegel's Asity *P. schlegeli*. The females of these two are very alike, mainly dull olive. The males resemble one another in having bare bright blue skin on the sides of the head in the breeding season. But whereas the male of the eastern species is predominantly black, that of the Sambirano is mainly olive and yellow, only the head being black. Both are short billed and feed on fruit. Although they are confined to dense forest, with patience one can obtain a view. Nor are the two sunbird asities easy to see. It was formerly believed that they belonged to the sunbirds (Nectariniidae), of which there are some 100 species in Africa and southeast Asia, but only 2 in Madagascar. Superficially they are quite unlike the asities, having long, curved beaks like the true sunbirds — an apparently similar adaptation in both for the extraction of nectar from flowers. They are also much smaller, length 10 cm as against 15 cm. However, in the breeding season the males develop bare blue skin on the sides of the head as in the asities. There is also anatomical evidence that these two pairs belong to the same family. One of the second pair has been observed frequently, but not so Salomonsen's Sunbird Asity *Neodrepanis hypoxantha*, to the extent that it has even been believed to be extinct, but this is unlikely. Although the forests of the humid east have been greatly reduced, there are still sizeable areas. Possibly it prefers the forest canopy, where it would be almost impossible to see.

Fig. 7.4 *Euryceros prevostii*, the helmet vanga (P. Randriamanantsoa, after A. Grandidier)

Fig. 7.5 *Falculea palliata*, the sicklebilled vanga (P. Randriamanantsoa, after A. Grandidier)

Fig. 7.6 *Xenopirostris xenopirostris*, Lafresnaye's vanga (P.Randriamanantsoa, after A. Grandidier)

Fig. 7.7 *Leptoterus chabert*, Chabert's vanga (P. Randriamanantsoa, after A. Grandidier)

Fig. 7.8 *Leptoterus viridis*, white-headed vanga (P. Randriamanantsoa, after A. Grandidier)

The fifth endemic family, the vangas (Vangidae) are represented by 14 species only found in Madagascar with the minor exception that the Blue Vanga *Cyanolanius madagascarinus* is represented in the Comoros. It is only absent from the sub-desert, and is easily observed. The male in particular is strikingly beautiful — ultramarine above, white below. The vangas may be most closely related to the helmet shrikes (Prionopidae) of Africa, and have descended from a single ancestral stock which crossed the Mozambique Channel. If so, the subsequent radiation into 14 species is remarkable. It could be correlated with isolation in a strong habitat diversity under locally contrasting climatic regimes. All the vangas subsist largely on insects, and most have strong, hooked bills. This has its culmination in the strikingly heavy bill of the Helmet Bird *Euryceros prevostii*, confined to forest in the northeast. The Sicklebill *Falculea palliata* is exceptional in having developed a long, decurved beak, adapted for extracting insects from holes in trees, much as do the wood hoopoes (Phoeniculidae) of Africa, quite unrelated, but with a similar bill. Living as it does mainly in the western savanna and subdesert, where the country is relatively open, the Sicklebill is not difficult to see, especially as it is a conspicuous black and white. Pollen's Vanga, Van Dam's Vanga and Lafresnaye's Vanga (*Xenopirostris polleni*, *damii*, *xenopirostris*) are all very similar, particularly in their heavy bills, and are an example of closely-allied species replacing one another — occurring respectively in the humid east, western savanna and subdesert. Pollen's Vanga is extremely similar in colour to the Tylas Vanga *Tylas eduardi*, only differing markedly in its heavier bill, the two apparently living alongside one another in dense forest. While colour mimicry is well known in insects, such instances are less common in birds. Pollen's Vanga, with its heavy bill, might be the better able to defend itself against an enemy, and so be the model. Colour mimicry could thus be of advantage to the Tylas Vanga. Chabert's Vanga *Leptoterus chabert*, found throughout, is markedly gregarious,

Fig. 7.9 *Vanga curvirostris*, hook-billed vanga (P. Randriamanantsoa, after A. Grandidier)

Fig. 7.10 *Hypositta corallirostris*, the coral-billed nuthatch vanga (P. Randriamanantsoa, after A. Grandidier)

a habit perhaps retained from prionopid ancestors. The Coral-billed Nuthatch Vanga *Hypositta corallirostris*, confined to the humid east, finds its food by climbing, much in the manner of a nuthatch (*Sitta*), a common enough bird in western Europe for example. Nevertheless, it may represent an extreme of vangid evolution. Unlike any true nuthatch, it is only known to climb upwards, never sideways or downwards.

Although the couas are usually regarded as belonging to the cuckoos (Cuculidae), the ten species form a compact and distinctive genus confined to Madagascar, of uncertain ancestry. At least they form a subfamily, and it is appropriate to consider them here. Like the Vangas, they illustrate what can be achieved by evolution through habitat diversity. Unlike them, however, they show little variation in bill shape, while all when adult have a patch of bare skin around the eyes, mainly bright blue. But there is much variation in size and colour of plumage. The smallest is Verreaux's Coua *Coua verreauxi*, only 38 cm long, the largest the Giant Coua *C. gigas*, measuring 62 cm. The apparently extinct Delalande's Coua *C. delalandei* was nearly as large, measuring 57 cm. It is said to have lived entirely on molluscs, broken open on rocks. The others apparently all subsist on a mixture of fruit and insects. Excluding Delalande's Coua, the forests of the humid east hold 3 species, the Blue Coua, Reynaud's Coua and the Red-breasted Coua *Coua caerulea*, *reynaudii*, and *serriana*. The first of these is an arboreal feeder, the other two terrestrial, Reynaud's Coua feeding mainly on insects rather than fruit, the Red-breasted Coua the reverse. Thus competition for food between these three seems to be largely avoided (nor would the mollusc specialist Delalande's Coua have been a competitor). For the western savanna and subdesert, it must suffice to say that there is one arboreal feeder predominant in each, respectively the Crested Coua *C. cristata* and Verreaux's Coua. Of the terrestrial ones, the Giant Coua and Coquerel's Coua *C. coquereli* live alongside one another in the western savanna and are almost identical in colour. But the latter is much the smaller, measuring 42 as against 62 cm, so that they probably take food of different sizes. Not all the species in the large, almost cosmopolitan cuckoo family are parasitic breeders. Indeed, every coua makes its own nest, an open cup of twigs, with usually 2 eggs (plain white). It is always above the ground, probably thereby reducing danger from predators.

OTHER BIRDS (EXCEPTING MIGRANTS AND SEA BIRDS)

Some numerical comparisons with Zambia are appropriate, since Madagascar lies in southcentral Africa at approximately the same latitudes, and is of comparable area. For this purpose, Benson *et al.* (1971) has been used. For Madagascar, any species in the endemic families discussed above are included. Even so, Madagascar has a total of only 189 compared to Zambia's 620. Subdivisions can be made into non-passerines — water birds, birds of prey (raptors), ground dwellers, others — and passerines (song birds) as subtitled below.

Water birds

Madagascar is relatively rich in water birds (grebes, herons, ibises, ducks, rails, plovers) — 58 species as compared to 87, yet in proportion to the totals 30% compared to 14%. Colonization by such birds across the Mozambique Channel may be relatively easy, perhaps assisted by rafts of floating vegetation brought down by rivers in spate. Some of these water birds are identical in Madagascar and Africa generally — for example the Little Grebe *Podiceps ruficollis*, Buff-backed Heron or Cattle Egret *Bubulcus ibis*, Dwarf Goose *Nettapus auritus*, and Red-knobbed Coot *Fulica cristata*. Two such others are the Gray-headed Gull *Larus cirrocephalus*, and Whiskered Tern *Chlidonias hybrida* (family gulls and terns Laridae), which both breed and feed inland. However, for other members of this family, see under sea birds below. Nevertheless, there are some very striking water bird endemics, whose conservation is important but may pose special problems. The wide ranging Little Grebe seems to be increasing at the expense of the endemic Madagascar Little Grebe *P. pelzelini*. This may be due to the introduction of herbivorous exotic fish, thereby changing the structure of many lakes and ponds, and modifying food sources in favour of the Little Grebe. The two species can mate and produce young, so that the endemic one may be endangered on this count too. On the outskirts of Tananarive, as in the heronry at Tsimbazaza, both the Madagascar Squacco Heron *Ardeola idae* and Squacco Heron *A. ralloides* were represented (at least in 1973). The former is easily distinguishable in the breeding season by its wholly white plumage. It is one of a few species which spend the non-breeding season in Africa, in this particular case perhaps in Zaire. As a breeder it is only otherwise known from Aldabra, where it may have become recently established from "fall-outs" on passage.

The Crested Wood Ibis *Lophotibis cristata* is in its own special genus. In general colour — chestnut, with white in the wings — it contrasts with the much duller Green Ibis *Bostrychia olivacea* of tropical Africa. But this may be its nearest relative, the ancestral stock having crossed the Mozambique Channel perhaps more than a million years ago. Unfortunately it is a popular article of diet. One practice is to wait until the young are about ready to leave the nest, and as large as possible, when they are taken, with the prospect also of snaring a parent (or with luck, both). By law the bird is protected, but it is normal for this to be disregarded. In some areas where it was formerly numerous, it may have disappeared completely. There are three endemic species of ducks (Anatidae). Meller's Duck *Anas melleri* is confined to the humid east, and represents the Mallard *A. platyrhynchos*, so familiar in Europe, and the Yellow-

Fig. 7.11 *Podiceps pelzelini*, Madagascar little grebe (P. Randriamanantsoa, after A. Grandidier)

Fig. 7.12 *Lophotibis cristata*, the crested wood ibis (B. Ranjato, after A. Grandidier)

Fig. 7.13 *Anas melleri*, Meller's duck (P. Randriamanantsoa, after A. Grandidier)

Fig. 7.14 *Dryolimnas cuvieri*, the white-throated rail (P. Randriamanantsoa, after A. Grandidier)

billed Duck *A. undulata*, of Africa. Bernier's Teal *A. bernieri* is only known from a few marshes in the west, and merits special protection. Its nearest relative is not in Africa, but is the Grey Teal *A. gibberifrons*, of southeast Asia and Australia. The Madagascar Pochard *Aythya innotata* is perhaps now confined to Lake Alaotra, and obviously requires attention too. Its nearest relative is the White-eyed Pochard *A. nyroca*, widespread as a breeder in Eurasia, but unknown to breed in tropical Africa. Five of the 11 species of rails (Rallidae) are endemics. A sixth is the White-throated Rail *Dryolimnas cuvieri*, the most distinctive of all and of uncertain ancestry. It can be found in almost any marshy area, and is only otherwise known from the Aldabra archipelago, 400 km to the northwest of the northern tip of Madagascar, and where (as so often happens to rail populations colonizing small islands) it has become flightless. Even the Moorhen *Gallinula chloropus*, so much in evidence in western Europe, and widespread in Africa, has developed a local peculiarity. The undercoverts of the tail, instead of being the normal white colour (constantly in evidence through flicking of the tail) are washed with rufous. Milon *et al.* place the Madagascar Jacana *Actophilornis albinucha* in the family Rallidae, but it belongs to the Jacanidae. It is another example of an endemic species, its nearest relative the African Jacana *A. africanus.* In both species the male is the smaller, and both are probably polyandrous; two males associated with one female, she taking no part in the incubation of the eggs. Another very distinct endemic is the Madagascar Snipe *Gallinago macrodactyla*, confined to the humid east. It is much larger and heavier than the Common Snipe *G. gallinago* of Europe or the Ethiopian Snipe *G. nigripennis*. It has the same noisy display flight in which the feathers of the tail (not wings as in Milon *et al.*) are vibrated, but louder. Its status needs watching. Undue disturbance of its habitat for rice growing could cause its extinction.

Birds of prey (raptors)

These comprise 20 species of eagles, goshawks, falcons, and owls, compared with 54 in Zambia. There are no vultures, presumably because there are no large mammals (such as antelopes) to provide carrion. There are only two large eagles. The Madagascar Fish Eagle *Haliaeetus vociferoides* is the ecological equivalent of the African Fish Eagle *H. vocifer* of Africa. The Madagascar Serpent Eagle *Eutriorchis astur*, confined to forest in the humid east, and little known, is the other. One might think that the lemurs would support some large eagle, although there is no evidence that this one ever takes them. It represents the only endemic raptor genus, although there are a number of peculiar species. Henst's Goshawk *Accipiter henstii* is the representative of the Black Goshawk *A. melanoleucus* of Africa and the Goshawk *A. gentilis*, which breeds in northern temperate climes, and to which Henst's Goshawk is most similar in appearance. The adults of both are barred below, but that of the African species plain white. The Madagascar Cuckoo Hawk *Aviceda madagascariensis* is probably derived from the African Cuckoo Hawk *A. cuculoides* (the nestlings of this hawk and Henst's Goshawk are eaten, the parents being snared at the nest as well if possible, so here is a problem not confined to the Crested Wood Ibis). The Madagascar Kestrel *Falco newtoni* only occurs regularly outside Madagascar on Aldabra. It is probably derived from the larger Rock Kestrel *F. tinnunculus*, so widespread in Eurasia and Africa. It is an attractive little bird, one of the few endemics which may have increased due to man's activities rather than the reverse. Cultivation would have favoured an increase in rodents, on which it thrives as well as on insects. Also, it commonly lays its eggs and rears its young in crevices in buildings. It is a familiar sight by the larger highways, perched on telegraph poles. The Madagascar Buzzard *Buteo brachypterus*, another endemic species, is still plentiful. In colour it is extraordinarily similar to the Madagascar Cuckoo Falcon, not distinguishable at a distance except by its mode of flight. Thus it is much more of a soarer. Among non-endemics, the Bat-eating Hawk *Macheiramphus alcinus* is identical with the African bird. It subsists largely on bats, and is therefore crepuscular, so it escapes notice, and may be more common than is supposed. As bats often adapt to man's dwellings, and thereby increase and this might benefit the bird. The Yellow-billed

Fig. 7.15 *Haliaetus vociferoides*, The Madagascar fish eagle (P. Randriamanantsoa, after A. Grandidier)

Kite *Milvus migrans*, also identical with the African bird, scavenges around villages, and must be increasing. The Peregrine *Falco peregrinus*, practically cosmopolitan even as a breeder, exists in Madagascar, but seems to be uncommon.

Of the 6 owls, the Barn Owl *Tyto alba* and the Marsh Owl *Asio capensis* belong to the same species as in Africa, the former almost cosmopolitan. The Madagascar Long-eared Owl *Asio madagascariensis* is related to the Long-eared Owl *A. otus*, widespread in northern temperate climates, and with outlying populations in the more mountainous parts of northeastern Africa. The Madagascar Scops Owl *Otus rutilus* and the White-browed Owl *Ninox superciliaris* have an Asiatic, not African, ancestry. Soumagne's Owl *Tyto soumagnii* is of uncertain ancestry, and altogether enigmatic. It is only known by a few records from dense forest in the humid east, and is said to subsist on frogs. Like other owls, it probably has a distinctive call, which it would be interesting to identify. This could lead to some determination of its status.

Ground dwellers

These consist of quails, guinea fowls, mesites, buttonquails, pratincoles and sandgrouse, of which there are only 9 species in all, compared to 30 in Zambia. The 3 mesites have been considered under endemic families. The most conspicuous absence is of bustards (Otididae) of which in southern and eastern Africa there are some 10 species. Nor are there any dikkops (Burhinidae), otherwise pantropical. It is surprising that neither of these families has gained any access to the more open parts of the west. The

Fig. 7.16 *Asio madagascariensis*, the Madagascar long-eared owl (P. Randriamanantsoa, after A. Grandidier)

Fig. 7.17 *Margaroperdrix madagascariensis*, the Madagascar partridge (P. Randriamanantsoa, after A. Grandidier)

Madagascar Partridge *Margaroperdix madagascariensis* may be misnamed, even though it is about the size of a partridge. It has been suggested recently, on the basis of the pattern of downy chicks, that it has a quail ancestry, and that subsequently it increased in size. To be sure, it is by now so distinct as to belong to an endemic genus. It occurs throughout, except in dense forest, but is diminishing through hunting and poaching. The Common Quail *Coturnix coturnix* and the Harlequin Quail *C. delegorguei* do not differ from these two species in Africa, and must have become established far more recently. The Common Quail frequents the central highlands, and its call sounds exactly the same as in South Africa (or indeed western Europe — for example). The Harlequin Quail is only known from the west, in relatively dry, low lying areas. The Helmeted Guineafowl *Numida mitrata* is widespread, and is exactly the same as the birds of coastal eastern Africa. It could only have been introduced by man. Probably it is overhunted for food, but its disappearance would not be such a tragedy as that of the Madagascar Partridge. The Madagascar Buttonquail *Turnix nigricollis* is an endemic species, its nearest relative perhaps the extreme southern African form of the Hottentot Buttonquail *T. h. hottentota*. The pratincoles (Glareolidae) are highly migratory, so that one or other of the two southern African species might have been expected to have remained unchanged, with constant movement across the Mozambique Channel. But not so, for the Madagascar Pratincole *Glareola ocularis* is a very distinct species, though nearest to the Rock Pratincole *G. nuchalis* of Africa. This is another example of a species which (like the Madagascar Squacco Heron) spends the off-season in Africa, mostly in coastal Tanzania and Kenya. The Madagascar Sandgrouse *Pterocles personatus* is another endemic, confined to the west, probably derived from the Yellow-throated Sandgrouse *P. gutturalis*, of southern and eastern Africa.

Other non-passerines

There are 37 species, compared to 116 in Zambia, and of which those in endemic families have already been discussed; that is, courols (1) and ground rollers (5), and also the couas, as a subfamily of the cuckoos (10). Conspicuous by their absence are such families as the trogons (Trogonidae), hornbills (Bucerotidae), barbets (Capitonidae) and woodpeckers (Picidae). There seem to be various niches awaiting occupation. Thus in Zambia these families are represented by altogether 30 species, occupying a variety of habitats, from dense forest to dry savanna. All of them nest in holes in trees (available through decay), seemingly little used in Madagascar. Of the 4 Columbidae, the most beautiful is the Madagascar Blue Pigeon *Alectroenas madagascariensis*. It belongs to a genus only otherwise now represented in the Comoros, Aldabra and the Seychelles, its nearest relatives in southeast Asia. The Madagascar species is confined mainly to the humid east. It is confiding and excellent eating, so that concern may be felt for its future. The Mauritian species became extinct about 1830. Some form also existed on Farquhar and Providence, between Madagascar and the Seychelles, but has not been seen since 1822. The Green Pigeon *Vinago australis* and Long-tailed Dove *Oena capensis* belong to species also found in Africa, but the relationship of the Madagascar Turtle Dove *Streptopelia picturata* is less clear. Of the 3 species of Psittacidae, the Gray-headed Lovebird *Agapornis cana* belongs to an African genus. The Greater Vasa Parrot *Coracopsis vasa* and Lesser Vasa Parrot *C. nigra* are both wholly dark brown and differ in little but size (50 cm as against 40 cm), occurring alongside one another. Their mode of flight — slow wing beats with intermittent gliding — is unlike the rapid, continual wing beats of any African parrot, and they may be of ancient Asiatic ancestry. Both have colonized the two outer Comoros, and the smaller one is found in the Seychelles, uniquely only on the island of Praslin. The Lesser Cuckoo *Cuculus poliocephalus* is only about two thirds of the size of the Cuckoo *C. canorus* of Eurasia, but similar in colour. It is present from about September to March, spending the rest of the year nearer the equator, in Africa. It has a monotonous call, "ko-ko kof", to be heard at night as well as in the daytime. Like its northern relative, it is, of course, parasitic in its breeding; its favourite host the Madagascar Cisticola

Fig. 7.18 *Alectroenas madagascariensis*, the Madagascar blue pigeon (P. Randriamanantsoa, after A. Grandidier)

Fig. 7.19 *Agapornis cana*, the grey-headed lovebird (P. Randriamanantsoa, after A. Grandidier)

Fig. 7.20 *Ispidina madagascariensis*, Madagascar hunting kingfisher (B. Ranjato, after A. Grandidier)

Cisticola cherina. It is closely related to another form (the two are usually regarded as conspecific) which breeds in Asia, and also spends the off-season in Africa. Yet no form breeds in Africa. The Thick-billed Cuckoo *Pachycoccyx audeberti* is also parasitic, and only differs from the African form in being slightly larger in size. In Madagascar it has not been recorded since 1922. As in Africa it is probably very uncommon, but not necessarily extinct. The non-parasitic Madagascar Coucal *Centropus toulou* is an endemic species (except that it has colonized the Aldabra archipelago), although superficially similar to others in Africa and Asia. It is widespread and common throughout.

The Madagascar Nightjar *Caprimulgus madagascariensis*, common throughout, has even been heard singing from roofs in the middle of Tananarive. Comparison of skins (and no less significantly, of tape recordings of the song) indicate that it may be conspecific with the Jungle Nightjar *C. asiaticus*. The Collared Nightjar *C. enarratus* is confined to the forests of the humid east and the Sambirano, its ancestry quite uncertain. The eggs have only been found once for certain. Highly peculiarly, they were laid above the ground, and were unmarked. There are 4 species of swift (Apodidae), probably all of African origin. The White-bellied (or Alpine) Swift *Apus melba* even breeds in southern Europe, as far north as Switzerland. There are only 2 species of kingfisher (Alcedinidae), compared to 12 in Zambia. The Madagascar Malachite Kingfisher *Alcedo vintsioides* is common and conspicuous; unlike its African relative, the Malachite Kingfisher *A. cristata*, not always near water. There is only the one bee-eater (Meropidae), the Madagascar (or Olive) Bee-eater *Merops superciliosus*, yet 8 in Zambia. It is often stated that the local birds, which lay eggs in September and October, migrate to Africa, where in coastal Mozambique identical birds breed in these months. Such a movement could only be proved by recovery of birds which had been ringed. On the other hand, the Broad-billed (or Cinnamon) Roller *Eurystomus glaucurus*, which lays mainly in November, spends the off-season in Zaire, with passage records from the Comoros, Aldabra, Malawi and eastern Zambia. It is larger than any African form of the same

Fig. 7.21 *Motacilla flaviventris*, Madagascar wagtail (P. Randriamanantsoa, after A. Grandidier)

species. The Madagascar race of the Hoopoe *Upupa epops* is a most remarkable bird, its call quite different from that of this species in southern Eurasia and Africa. It differs from southern African birds not only in certain points of colour but in being larger and heavier. Probably it is another example of an Asiatic ancestry.

Passerines (song birds)

There are only 65 species, compared to 329 in Zambia. The proportion to the total is a mere 34% as against 53%, the African figure thus strikingly higher. The endemic families Philepittidae and Vangidae, providing 18 of the 65 species, have been discussed above. Excluding the Indian Myna *Acridotheres tristis* (originally confined to India), introduced by man to the east coast about 100 years ago (fortunately still not unduly widespread), and to Natal about the same time, there are only 3 species shared with Africa: the Brown-throated Sand Martin *Riparia paludicola*, the Stonechat *Saxicola torquata* and the Pied Crow *Corvus albus*. All three can be seen when travelling by automobile along highways. The first, confined to the humid east, often nests in tunnels into road banks. The second is most plentiful in the humid east; and often in evidence, perched at the top of a bush. The same species is even found in Eurasia as well as Africa. The third occurs identically throughout sub-Saharan Africa and Madagascar. It soars and rides on up-draughts, so that crossings of the Mozambique Channel may be frequent. It is commensal with man as a scavenger, and is undoubtedly on the increase.

The Madagascar Bush Lark *Mirafra hova* is the only species of this family, compared to 11 in Zambia and many more in drier areas to the southward. Yet there is much suitable open ground in the west, and some of the African larks have regular movements. The Mascarene Martin *Phedina borbonica* is another species which spends the off-season in Africa (in coastal areas). This and the Brown-throated

Sand Martin *Riparia paludicola* are the only 2 members of the family which breed locally, yet there are 17 in Zambia. To cross the Mozambique Channel should present no problem for such mobile birds, and Madagascar ought to be able to support more species. The Crested Drongo *Dicrurus forficatus* is widespread, and has also colonized Anjouan in the Comoros. It is a close relative of the Fork-tailed Drongo *D. adsimilis*, of Africa, but has a conspicuous tuft of feathers on the forehead. The Madagascar Wagtail *Motacilla flaviventris* is one of the few examples of an endemic which may have benefited from man. It is frequent around human dwellings, nesting in them, its nearest relative probably the Cape Wagtail *M. capensis*, of southern and eastern Africa. There is not a single pipit, yet 7 breed in Zambia. The Madagascar Cuckoo Shrike *Coracina cinerea* is the only representative of its family, probably of Asiatic ancestry. So too, without question, is the Madagascar Bulbul *Hypsipetes madagascariensis*. Differences in plumage from certain Indian forms are but slight. Yet 5 other species, commonly accepted as of this same family (Pycnonotidae), which Milon *et al.* (pp.197 – 201) place in the babblers (Timaliidae), seem to be of African origin. One, the Dusky Tetraka *Phyllastrephus tenebrosus*, they place under the same heading as the Tetraka *P. madagascariensis*. It is certainly a valid species, only known from the Sianaka forest, its nearest ally the Short-billed Tetraka *P. zosterops*. The discovery of one of these 5 was announced as recently as 1972, from dry forest in the southwest: Appert's Tetraka *P. apperti*.

What species to place in the family Timaliidae is always debatable, but perhaps in Madagascar there are 7. There are 4 small jeries (*Neomixis*, *Hartertula*), arboreal rather than terrestrial feeders. The remaining 3, Crossley's Babbler *Mystacornis crossleyi*, the White-throated Oxylabes *Oxylabes madagascariensis* and Yellow-browed Oxylabes *Crossleyia xanthophrys*, each belongs to its own special genus confined to the humid east, all apparently feeding on or near the floor of dense forest. The juvenile

Fig. 7.22 *Oxylabes madagascariensis*, the white-throated oxylabes (P. Randriamanantsoa, after A. Grandidier)

of Crossley's Babbler is remarkable in having a predominantly rufous dress. In colour, the adult male resembles that of the White-bellied Antbird *Myrmeciza longipes* (family Formicariidae), of tropical South America, although on geographical grounds a relationship seems unlikely.

Of the 5 species in the family thrushes (Turdidae), the Stonechat has already been mentioned. The Madagascar Magpie Robin *Copsychus albospecularis*, like the Madagascar Bulbul, is of Asiatic origin. Both are represented by allied, large sized, forms in the intervening Seychelles. The robin has split into 3 well-marked subspecies, the 1 in the dry west with much more white in the plumage (in the male otherwise black) than the 2 in the humid east, white reduced to a minimum in the northeast, where the rainfall is highest. At least in the west it takes advantage of man. It can nest in holes in road banks. A pair has even been recorded using a hole in the wall of an hotel from which a brick was missing. Milon *et al.* (p.207) place 3 different species under the one heading, "Robin Chat" (sic) (*Pseudocossyphus*). They surely belong to the same genus (*Monticola*) as for example the Rockthrush *M. saxatilis*, which breeds in Eurasia. Most of the species inhabit relatively or even entirely open country, as in South Africa, so it is remarkable that the Eastern Madagascar Rockthrush *M. sharpei*, of the humid east, inhabits dense forest. Surely its ancestor (probably African) did not live in this enclosed environment. The subdesert Rockthrush *M. imerina* (Imerina geographicaly inept) inhabits sandy dunes covered with low shrubs interspersed with a species of *Euphorbia*. The genus is absent from the western savanna as a whole, except for the discovery of a third species, Benson's Rockthrush *M. bensoni*, announced in 1971, from the Massif de l'Isalo, in much more typical habitat — open rocky hillsides.

Of 7 species in the family warblers (Sylviidae), the Madagascar Swamp Warbler *Acrocephalus newtoni* and the Madagascar Cisticola *Cisticola cherina*, although they belong to endemic species, represent genera widespread in the Old World, the latter with 24 species in Zambia. The other 5 represent 4 endemic

Fig. 7.23 *Terpsiphone mutata*, Madagascar paradise flycatcher (P. Randriamanantsoa, after A. Grandidier)

Fig. 7.24 *Lonchura nana*, Madagascar mannikin (P. Randriamanantsoa, after A. Grandidier)

genera, of which one can claim no more than that they belong to the warblers. The Thamnornis *Thamnornis chloropetoides* is peculiar to the subdesert, Rand's Warbler *Randia pseudozosterops* to forest in the humid east. By contrast, the Tsikirity *Nesillas typica* occurs throughout, and has colonized the Comoros and Aldabra. The two species of emu-tail, about the same size, are confined to the humid east, and show strong ecological contrast, so avoiding competition with one another. The Brown Emu-tail *Dromaeocercus brunneus* lives on or near the floor of the forest, and is particularly difficult to see, with the Gray Emu-tail *D. seebohmi* in the undergrowth of marshes. The tail of both is highly characteristic: elongate, the lateral barbs feebly developed. When disturbed, they seem to prefer to escape by moving through the dense undergrowth rather than by flying, facilitated by the tail structure. Independently, unindicative of any relationship, this has also been evolved in another group, the emu wrens, genus *Stipiturus*, in faraway Australia.

In the family flycatchers (Muscicapidae), the Madagascar Paradise Flycatcher *Terpsiphone mutata* occurs throughout, and belongs to a genus found in Africa and southeast Asia. Ward's Flycatcher *Pseudobias wardi*, on the other hand, belongs to another endemic genus, confined to forest in the humid east. The plumage pattern of the juvenile is somewhat reminiscent of that in the flycatcher genus *Batis*, of Africa. The sexes are alike in *Pseudobias*, dissimilar in most species of *Batis* but not all. There might be a common origin, but if so a long way back in time. There is an enigma over the ancestry of the Newtonia flycatchers, genus *Newtonia*, of which there are 4 species under the one heading in Milon *et al.* (p.220). The Common Newtonia *N. brunneicauda* occurs throughout, Archbold's Newtonia *N. archboldi* only in the subdesert, the Dark Newtonia *N. amphichroa* only in the humid east (in the undergrowth of dense forest), while the Fanovana Newtonia *N. fanovanae* is known merely by a single specimen from the humid east. This fourth one is much paler in colour than the third. It may inhabit the forest canopy, where it would be almost impossible to see. Quite unaware, some of us may have had this bird a few metres above our heads in forest at Pe'rinet (Andasibe). Eventually it might be picked out by some peculiarity in voice. Possibly these 4 species have a link with the genus *Humblotia* of Grand Comoro, which perhaps had an earlier ancestry in Africa, still untraced.

The Madagascar Starling *Saroglossa aurata* provides one more example of an Asiatic ancestry, with no close relationship to any African species. There are only 2 species of sunbird, family Nectariniidae, yet 24 in Zambia. One is much larger and longer billed than the other, so that they probably do not compete for food. The closest ally of the smaller Souimanga Sunbird *Nectarinia souimanga* seems to be in Asia. The resemblance on the underparts between males of it and the Olive-backed Sunbird *N. jugularis* (subspecies *flammaxillaris*, of Burma and Malaysia) is startling. The Madagascar Green Sunbird *N. notata* probably has an African origin. The Madagascar White-eye *Zosterops maderaspatana*, as in most parts of Africa, is the only representative of its family. The mannikin family (Estrildidae) is represented by a single very distinctive species (compared to 27 in Zambia), the Madagascar Mannikin *Lonchura nana*. The weavers and fodies (Ploceidae) have 4 species (but 31 in Zambia). They include the Red Fody *Foudia madagascariensis* and Red Forest Fody *F. omissa*, a genus only otherwise known from smaller islands in the western Indian Ocean. The former is basically a western savanna species, and the male in his red breeding dress (worn mainly from December to May) is even evident from an automobile on a highway. The second is confined to forest in the humid east. Due to forest reduction (for agriculture and timber felling) the Red Fody is increasing in the humid east, but was probably originally absent. At Pe'rinet (Andasibe) it now hybridizes with the Red Forest Fody. If this continues, it could lead to the disappearance of the latter in a pure form.

MIGRANTS

A few species which breed in Madagascar and cross the Mozambique Channel to spend the off-season in Africa are mentioned above. Dealt with here are visitors unknown to breed. Some 20 species of shore birds (plovers, stints, godwits, curlews, sandpipers) visit southern Africa and Madagascar, mostly the coasts, as migrants mainly from northern Eurasia, some of them familiar to observers in western Europe, for example, while on passage. Of land birds in the more strict sense, in Africa there are many species which are visitors from breeding quarters to the northward, some in enormous numbers. But the only 2 which visit Madagascar regularly are Eleonora's Falcon *Falco eleonorae* and the Sooty Falcon *F. concolor*. They occur in tropical and/or southern Africa for the most part only on passage, and where they are largely replaced by such species as the Lesser Kestrel *F. naumanni* and Hobby *F. subbuteo*. Eleonora's Falcon breeds in the Mediterranean region, the Sooty Falcon mainly in northeast Africa. Both breed very late, so that they can feed their young on small passerine birds on southward migration, although when in Madagascar they seem to feed mainly on grasshoppers. The only other land bird migrants so far recorded are the Swallow *Hirundo rustica* and Sand Martin *Riparia riparia*. Both are merely occasional, whereas in Africa they occur in hundreds of millions.

SEA BIRDS

These include petrels (Hydrobatidae), shearwaters (Procellariidae), tropicbirds (Phaethontidae), boobies (Sulidae), gulls and terns (Laridae) and skuas (Stercorariidae), which as a whole feed only in marine waters or their shores, and only come to land proper to breed in coastal areas. One species each of gull and tern has been mentioned under freshwater birds. But there are 30 known from the coasts, where the following breed: Wedge-tailed Shearwater *Puffinus pacificus*, White-tailed Tropicbird *Phaethon lepturus*, Southern Black-backed Gull *Larus dominicanus*, Caspian Tern *Hydroprogne caspia*, Swift Tern *Sterna bergii*, Roseate Tern *S. dougallii*, Sooty Tern *S. fuscata*, Bridled Tern *S. anaethetus* and Noddy

Fig. 7.25 *Ploceus nelicourvi*, the nelicourvi weaver (Sonnerat, 1792)

Anous stolidus. Also, the Red-footed Booby *Sula sula* breeds at Europa (in the Mozambique Channel) and Tromelin (400 km off the northeast coast), while the Brown Booby *Sula leucogaster* has been found breeding as near as the Iles Mitsios, northeast of Nosy Bé. In general, sea birds are far more wide ranging than land birds. Thus the Noddy is practically pantropical. So in general they are not of the same endemic interest as land birds, including islands such as Madagascar. Thus, while one could only deplore the destruction of all the local nesting colonies of the Roseate Tern, it does breed in other parts of the world. It would not be so tragic as the extirpation of, say, the Crested Wood Ibis.

SUMMARY

Space has only permitted reference to certain species and special points. Although the variety of species is much less than in Africa, this is more than compensated for by the uniqueness of so many, and their conservation is of outstanding importance in a world context. An appreciable proportion has an Asiatic, not African, ancestry. Madagascar seems to suffer from more than its fair share of hunting of birds, with more of a traditional basis than in Africa, where at least in the past larger prey (elephants, antelopes) was available. But above all, habitat conservation is vital. The birds of the dense forests of the humid east and the Sambirano are absolutely dependent on an unmodified environment. They are even less able to withstand any change than those of the savanna and subdesert.

ACKNOWLEDGEMENTS

I am very grateful to my friend the late Dr. Paul Griveaud, and to M. Philippe Oberle', for their comments on the original French version.

REFERENCES

Journals

Appert, O. (1980) Erste farbaufnahmen der rachenzeichnung junger kuas von Madagaskar (Cuculi, Couinae). *Orn. Beobacht.* 77, 85 – 101.

Benson, C.W. (1980) Fifty years of ornithology in the Malagasy Faunal Region. *Bull. Br. Orn. Club* 100, 76 – 80.

Benson, C.W., Colebrook-Robjent, J.F.R. and Williams, A. (1976 – 77) Contribution a' l'ornithologie de Madagascar. *Oiseau Revue fr. Orn.* 46, 103 – 34, 209 – 42, 367 – 86; 47, 41 – 64, 167 – 91.

Benson, C.W. and Irwin, M.P.S. (1975) The systematic position of *Phyllastrephus orostruthus* and *Phyllastrephus xanthophrys*, two species incorrectly placed in the family Pycnonotidae (Aves). *Arnoldia* (*Rhod.*) 7 (17).

Cracraft, J. (1971) The relationship and evolution of the rollers: families Coraciidae, Brachypteraciidae, and Leptosomatidae. *Auk* 88, 723 – 52.

Dhondt, A. (1976) Une nidification de l'Engoulevent a' collier. *Oiseau Revue fr. Orn.* 46, 173 – 74.

Farkas, T. (1974) On the biology of *Monticola imerinus*. *Bull. Br. Orn. Club* 94, 165 – 70.

Frost, P.G.H. (1975) The systematic position of the Madagascan partridge *Margaroperdix madagascariensis*. *Bull. Br. Orn. Club* 95, 64 – 68.

Keith, S. (1980) Origins of the avifauna of the Malagasy Region. *Proc. IV Pan-Afr. Orn. Congr.* 99 – 108.

Polunin, N.V.C. (1979) *Sula leucogaster* and other species in the Iles Mitsios, Madagascar. *Bull. Br. Orn. Club* 99, 110 – 11.

Books

Benson, C.W., Brooke, R.K., Dowsett, R.J. and Irwin, M.P.S. (1971) *The Birds of Zambia*. Collins, London.

Jolly, A. (1980) *A World Like Our Own*. Yale University Press, New Haven and London.

Milon, P., Petter, J.J. and Randrianasolo, G. (1973) *Faune de Madagascar, 35. Oiseaux*. ORSTOM, Tananarive; CNRS, Paris.

APPENDIX

Birds Breeding in the Present Time in Madagascar

Non-breeding visitors, such as the European Swallow *Hirundo rustica* and various palearctic-breeding shore birds, are excluded. Any differences in the scientific or English nomenclature from Milon *et al.* (1973) are intended to reflect a stricter observance of current practice. Their sequence, however, has been followed almost entirely, although in some instances it is somewhat unorthodox. Thus the gulls and terns (Laridae) (Milon et al., 108 – 16) are usually placed between the pratincoles (Glareolidae) and sandgrouse (Pteroclididae). In the 4 distribution columns, it is hardly necessary to state that the lack of a sign indicates apparent absence. The fourth column is restricted to the "spiny desert" as in Jolly (1980: map facing xiii). "Marine" species cannot be suitably fitted into any of the columns: see in particular the variations in the gulls and terns, of which only two are freshwater breeders. Some species are apparently highly restricted in distribution. Thus (cf. Milon *et al.*, 220), one of the newtonias *Newtonia fanovanae* is only known in the humid east from Fanovana, east of Perinet. Further instances can be found in the same work. Endemic families are indicated as such in capitals. Endemic genera not in endemic families, and endemic species not in endemic genera, are prefixed by an asterisk. "Endemic" includes some families, genera or species represented outside Madagascar provided that they are not known to also breed outside the Malagasy Faunal Region (Madagascar and outlying islands, as in Benson, 1980). Such an example among families is provided by the courols (Leptosomidae), which breed in the Comoros as well as in Madagascar; among genera the swallow or martin *Phedina*, which also breeds on Mauritius and Reunion — it occurs in Africa too, but merely as a non-breeding visitor; and among species the coucal *Centropus toulou*, also resident on Aldabra (the scops owl *Otus rutilus* is not asterisked because what is generally accepted as the same species also resides on Pemba Island, off Africa). In all such extensions of "endemic", it is assumed (perhaps always correctly) that the taxon was evolved in Madagascar, by far the largest and most diverse island. It is not practicable to deal with subspecies, of which there are very many endemic to Madagascar. Thus the Hoopoe *Upupa epops* is a very well marked subspecies, perhaps even better regarded as a full species. The Crested Coua *Coua cristata* has proliferated into 4 subspecies. Also, the Tsikirity *Nesillas typica* (family warblers) is represented by 3, with 3 others in the Comoros, and 2 full species, 1 on Moheli in the Comoros, 1 on Aldabra.

Recommended further reading on the fascinating subject of colonization and endemism is the paper by Keith (1980).

APPENDIX

Page ref. in Milon *et al.* (1973)	Scientific name	English name	Malagasy name	Humid east	Sambi-rano	Western savanna	South western subdesert
		Grebes (Podicipedidae)					
23	*Podiceps *pelzelnii*	Madagascar Little Grebe	Vivy, Kiborano	o	o	o	o
24	*Podiceps ruficollis*	Little Grebe	Vivy	o	o	o	o
25	*Podiceps *rufolavatus*	Madagascar Red-necked Grebe	Vivy	o			
		Shearwaters (Procellariidae)					
27	*Puffinus pacificus*	Wedge-tailed Shearwater	Vorotomany		MARINE		
		Tropicbirds (Phaethontidae)					
29	*Phaethon lepturus*	White-tailed Tropicbird	Kafotsy		MARINE		
		Pelicans (Pelecanidae)					
31	*Pelecanus rufescens*	Pink-backed Pelican	—			o	
		Cormorants (Phalacrocoracidae)					
34	*Phalacrocorax africanus*	Long-tailed Cormorant	Fangalamotamboay, Kontonboay, Vadimboay	o	o	o	o
		Darters (Anhingidae)					
35	*Anhinga rufa*	African Darter	Fangalamotomboay, Kontonboay, Vadimboay	o	o	o	o
		Herons etc. (Ardeidae)					
38	*Ardea cinerea*	Gray Heron	Vano			o	o
39	*Ardea purpurea*	Purple Heron	Vano-be	o	o	o	o
40	*Ardea *humbloti*	Madagascar Heron	Vano Humblot			o	o
41	*Egretta alba*	Great White Egret	Vano Fotsy-be, Kilandry-be	o	o	o	o
42	*Egretta garzetta*	Little Egret	Vano Fotsy (white phase), Vano Mainty (dark phase)	o	o	o	o
43	*Egretta ardesiaca*	Black Egret	Lombokohoma	o	o	o	

Page ref. in Milon *et al.* (1973)	Scientific name	English name	Malagasy name	Humid east	Sambi-rano	Western savanna	South western subdesert
44	*Bubulcus ibis*	Buff-backed Heron or Cattle Egret	Borompotsy	o	o	o	o
46	*Ardeola ralloides*	Squacco Heron	Mpiandry-voditatatra	o	o	o	o
48	*Ardeola *idae*	Madagascar Squacco Heron	Mpiandry-voditatatra	o	o	o	o
49	*Nycticorax nycticorax*	Night Heron	Goaka	o	o	o	o
51	*Butorides striatus*	Green-backed Heron	Vanomainty	o	o	o	o
51	*Ixobrychus minutus*	Little Bittern	Vanomainty	o	o	o	o
		Hamerkops (Scopidae)					
54	*Scopus umbretta*	Hamerkop	Takatra	o	o	o	o
		Storks (Ciconiidae)					
55	*Ibis ibis*	Yellow-billed Stork	Mefo			o	o
56	*Anastomus lamelligerus*	Openbill	Famaky, Akora	o	o	o	o
		Ibises etc. (Threskiornithidae)					
58	*Plegadis falcinellus*	Glossy Ibis	Doaka, Famakisifotra	o	o	o	o
59	*Threskiornis aethiopicus*	Sacred Ibis	Voronosy, Manarasoy-fotsy			o	o
60	**Lophotibis cristata*	Crested Wood Ibis	Akon'-ala, Lampira	o	o	o	o
61	*Platalea alba*	African Spoonbill	Sotrovava			o	o
		Flamingos (Phoenicopteridae)					
63	*Phoenicopterus ruber*	Greater Flamingo	Samaky, Garega				o
65	*Phoeniconaias minor*	Lesser Flamingo	Samaky				o
		Ducks, geese etc. (Anatidae)					
66	*Dendrocygna bicolor*	Fulvous Tree Duck	Tahia	o			o
68	*Dendrocygna viduata*	White-faced Tree Duck	Tsiriry, Vivy	o	o	o	o
69	*Anas hottentota*	Hottentot Teal	Kizazaka	o	o	o	o
70	*Anas erythrorhynchos*	Red-billed Teal	Rahaka, Fotsielatra, Sadakely, Menamolotro	o	o	o	o
71	*Anus *melleri*	Meller's Duck	Angaka, Akaka	o			
72	*Anas *bernieri*	Bernier's Teal	—			o	
72	*Nyroca *innotata*	Madagascar Pochard	Fotsy maso	o			
73	*Nettapus auritus*	Dwarf Goose	Vorontsara, Tsaravanga, Fiamboa, Soa-fify	o	o	o	o
74	*Sarkidiornis melanotos*	Knob-billed Goose	Arosy, Tsivongo, Ongongo	o	o	o	o
75	*Thalassornis leuconotus*	White-backed Duck	Beloha, Maheri-loha, Danamona	o	o	o	o

Page ref. in Milon *et al.* (1973)	Scientific name	English name	Malagasy name	Humid east	Sambi-rano	Western savanna	South western subdesert
		Falcons, kestrels (Falconidae)					
76	*Falco peregrinus*	Peregrine	Voromahery	o	o	o	o
79	*Falco *newtoni*	Madagascar Kestrel	Hitsikitsika	o	o	o	o
80	*Falco *zoniventris*	Banded Kestrel	Fandrao-kibo	o	o	o	o
		Eagles, hawks etc. (Accipitridae)					
81	*Aviceda *madagascariensis*	Madagascar Cuckoo Hawk	Bubuka, Endrina	o	o	o	
82	*Milvus migrans*	Yellow-billed Kite	Papango	o	o	o	o
82	*Macheiramphus alcinus*	Bat-eating Hawk	Fandrantsandambo, Sendika	o	o	o	
83	**Eutriorchis astur*	Madagascar Serpent Eagle	—	o			
84	*Haliaeetus *vociferoides*	Madagascar Fish Eagle	Ankoay		o	o	
85	*Buteo *brachypterus*	Madagascar Buzzard	Bemano Beririna, Hindry	o	o	o	o
86	*Accipiter *madagascariensis*	Madagascar Sparrow Hawk	Firasa, Fandrao-kibo	o	o	o	o
87	*Accipiter *henstii*	Henst's Goshawk	Rehila, Firasa, Sendika	o	o	o	
88	*Accipiter *francesiae*	Madagascar Goshawk	Hitsikitsik'ala	o	o	o	o
89	*Polyboroides *radiatus*	Madagascar Harrier Hawk or Gymnogene	Fihiaka, Fanindry	o	o	o	o
90	*Circus *maillardi*	Maillard's Harrier	Fandraslambo	o	o	o	o
		Partridges, quails, guineafowls (Phasianidae)					
91	**Margaroperdix madagascariensis*	Madagascar Partridge	Tsipoy, Traotrao	o	o	o	
92	*Coturnix delegorguei*	Harlequin Quail	Kibon'omby			o	o
93	*Cortunix coturnix*	Common Quail	Papelika	o			
93	*Numida mitrata*	Helmeted Guineafowl	Akanga	o	o	o	o
		Mesites (Mesitornithidae) (ENDEMIC FAMILY)					
95	*Mesites variegata*	White-breasted Mesite	Tolohon'ala, Fangadekovy			o	
96	*Mesites unicolor*	Brown Mesite	Roatelo	o			
96	*Monias benschi*	Bensch's Rail	Naka				o
		Buttonquails (Turnicidae)					
98	*Turnix *nigricollis*	Madagascar Buttonquail	Kibo, Kibobo	o	o	o	o

Page ref. in Milon *et al.* (1973)	Scientific name	English name	Malagasy name	Humid east	Sambi-rano	Western savanna	South western subdesert
		Rails etc. (Rallidae)					
100	*Canirallus *kioloides*	Gray-throated Rail or Wood Rail	Tsikoza-ala Tsikoza Vohitra, Drovikala		o	o	
101	**Dryolimnas cuvieri*	White-throated Rail	Tsikoza	o	o	o	
102	*Rallus *madagascariensis*	Madagascar Rail	Tsikoza	o			
102	*Porzana pusilla*	Baillon's Crake	Tsikoza	o	o	o	o
103	*Amaurornis *olivieri*	Olivier's Rail	Tsikoza			o	
103	*Sarothrura *insularis*	Madagascar Flufftail	Biry, Tsobeboka	o	o		
104	*Sarothrura *watersi*	Waters's Flufftail	—	o			
105	*Gallinula chloropus*	Moorhen	Aretaka	o	o	o	o
105	*Porphyrio alleni*	Allen's Reedhen	Talevakely, Hesetrika	o	o	o	
106	*Porphyrio madagascariensis*	King Reedhen	Talevana	o	o	o	o
107	*Fulica cristata*	Red-knobbed Coot	Tsohia, Otrika, Vontsiona, Akoharano	o	o	o	o
		Jacanas (Jacanidae)					
99	*Actophilornis *albinucha*	Madagascar Jacana	Tsikay	o	o	o	o
		Gulls, terns (Laridae)					
108	*Larus dominicanus*	Southern Black-backed (or Kelp) Gull	Kolokoloka		MARINE		
109	*Larus cirrocephalus*	Gray-headed Gull	Varevake	o		o	o
109	*Hydroprogne caspia*	Caspian Tern	Samby		MARINE		
110	*Chlidonias hybrida*	Whiskered Tern	Samby	o		o	
113	*Sterna dougallii*	Roseate Tern	Samby, Kirinina		MARINE		
113	*Sterna bergii*	Swift Tern	Samby		MARINE		
114	*Sterna fuscata*	Sooty Tern	Samby, Sarevaka		MARINE		
115	*Sterna anaethetus*	Bridled Tern	Samby, Mavolambosy		MARINE		
116	*Anous stolidus*	Noddy	—		MARINE		
		Plovers (Charadriidae)					
119	*Charadrius pecuarius*	Kittlitz's Sandplover	Viky-viky	o	o	o	o
122	*Charadrius *thoracicus*	Black-fronted Sandplover	Viky-viky	o			o
121	*Charadrius marginatus*	White-fronted Sandplover	Viky-viky, Fandy-fasky	o	o	o	o
123	*Charadrius tricollaris*	Three-banded Plover	Vorombato	o	o	o	o
		Snipes (Scolopacidae)					
133	*Gallinago *macrodactyla*	Madagascar Snipe	Kitanatana, Harakaraka, Rava-rava	o			

MAD-K

Page ref. in Milon *et al.* (1973)	Scientific name	English name	Malagasy name	Humid east	Sambi-rano	Western savanna	South western subdesert
		Painted Snipes (Rostratulidae)					
134	*Rostratula benghalensis*	Painted Snipe	Takoka	o	o	o	o
		Stilts (Recurvirostridae)					
136	*Himantopus himantopus*	Black-winged Stilt	Takapaly, Tsakaranta	o	o	o	o
		Pratincoles (Glareolidae)					
137	*Glareola *ocularis*	Madagascar Pratincole	Viko-viko	o	o	o	o
		Sandgrouse (Pteroclididae)					
139	*Pterocles *personatus*	Madagascar Sandgrouse	Katrakatra			o	o
		Pigeons, doves (Columbidae)					
140	**Alectroenas madagascariensis*	Madagascar Blue Pigeon	Finingo-manga, Finingo-menamaso, Finingo-menavody	o	o	o	
141	*Treron australis*	Green Pigeon	Finingo-maitso, Voron'adabo	o	o	o	o
142	*Streptopelia *picturata*	Madagascar Turtle Dove	Domohina	o	o	o	o
142	*Oena capensis*	Long-tailed Dove	Katoto, Tsiazotenonina	o	o	o	o
		Parrots (Psittacidae)					
144	*Coracopsis vasa*	Greater Vasa Parrot	Siotsabe, Boloky, Boeza-be	o	o	o	o
145	**Coracopsis nigra*	Lesser Vasa Parrot	Koakio, Boloky	o	o	o	o
146	*Agapornis *cana*	Gray-headed Lovebird	Sarivazo, Karaoka, Kitrehoka	o	o	o	o
		Cuckoos, coucals, couas (Cuculidae)					
149	*Cuculus poliocephalus*	Lesser Cuckoo	Kakafotra, Taotao-kafo	o	o	o	o
150	*Pachycoccyx audeberti*	Thick-billed Cuckoo	—	o			
151	*Centropus *toulou*	Madagascar Coucal	Toloho	o	o	o	o
152	*Coua caerulea*	Blue Coua	Taitso-manga, Mariha	o	o		
153	*Coua cristata*	Crested Coua	Tivoka, Tsiloko	o	o	o	o
155	*Coua verreauxi*	Verreaux's Coua	Tivoka				o
155	*Coua reynaudii*	Reynaud's Coua	Koa, Taitoaka	o	o		
156	*Coua serriana*	Red-breasted Coua	Koa, Tsivoka	o			

Page ref. in Milon *et al.* (1973)	Scientific name	English name	Malagasy name	Humid east	Sambi-rano	Western savanna	South western subdesert
157	**Coua delalandei*	Delalande's Coua	Famakiakora	o			
158	*Coua gigas*	Giant Coua	Eoke			o	o
158	*Coua ruficeps*	Red-capped Coua	Aliotsy, Akoke			o	o
160	*Coua cursor*	Running Coua	Aliotsy				o
161	*Coua coquereli*	Coquerel's Coua	Gory, Aliotsy		o	o	
		Owls (Strigidae)					
162	*Asio capensis*	Marsh Owl	Vorombozaka, Vorondolo	o	o	o	o
163	*Asio *madagascariensis*	Madagascar Long-eared Owl	Vorondolo, Hanka	o	o	o	
163	*Ninox *superciliaris*	White-browed Owl	Vorondolo			o	o
164	*Otus rutilus*	Madagascar Scops Owl	Toro-toroka	o	o	o	
		Barn Owls (Tytonidae)					
165	*Tyto alba*	Barn Owl	Tararaka, Hora	o	o	o	o
166	*Tyto *soumagnii*	Soumagne's Owl	—	o			
		Nightjars (Caprimulgidae)					
167	*Caprimulgus *madagascariensis*	Madagascar Nightjar	Langoapaka, Tantarako, Matoriandra	o	o	o	o
168	*Caprimulgus *enarratus*	Collared Nightjar	Matoriandra	o	o		
		Swifts, spinetails (Apodidae)					
169	*Apus barbatus*	African Black Swift	Tsiditsidina	o	o	o	o
169	*Apus melba*	White-bellied (or Alpine) Swift	Hela-kela	o	o	o	o
170	*Cypsiurus parvus*	African Palm Swift	Manaviandro, Tsiditsidina	o	o	o	o
171	*Zoonavena *grandidieri*	Madagascar Spinetail	Manaviandro, Tsiditsidina	o	o	o	o
		Rollers (Coraciidae)					
172	*Eurystomus glaucurus*	Broad-billed (or Cinnamon) Roller	Vorom-baratra, Tsararaka, Hahak	o	o	o	o
		Courols (Leptosomidae) (ENDEMIC FAMILY)					
172	*Leptosomus discolor*	Courol	Vorondreo, Kirombo	o	o	o	o
		Ground Rollers (Brachypteraciidae) (ENDEMIC FAMILY)					
174	*Brachypteracias leptosomus*	Short-legged Ground Roller	Fandikalalana	o			

Page ref. in Milon *et al.* (1973)	Scientific name	English name	Malagasy name	Humid east	Sambi-rano	Western savanna	South western subdesert
174	*Brachypteracias squamigera*	Scaled Ground Roller	Roatelo	o			
175	*Atelornis pittoides*	Pitta-like Ground Roller	Fangadiovy, Tsakoka	o	o		
176	*Atelornis crossleyi*	Crossley's Ground Roller	—	o			
176	*Atelornis chimaera*	Long-tailed Ground Roller	Bokitsy, Teroboky, Tolohoranto				o
		Bee-eaters (Meropidae)					
178	*Merops superciliosus*	Madagascar (or Olive) Bee-eater	Kirio, Kirioka	o	o	o	o
		Kingfishers (Alcedinidae)					
179	*Alcedo *vintsioides*	Madagascar Malachite Kingfisher	Vintsy	o	o	o	
180	*Ispidina *madagascariensis*	Madagascar Pygmy Kingfisher	Vintsy-ala, vintsy-mena	o	o	o	
		Hoopoes (Upupidae)					
181	*Upupa epops*	Hoopoe	Takodara	o	o	o	o
		Asities, sunbird asites (Philepittidae) (ENDEMIC FAMILY)					
182	*Philepitta castanea*	Velvet Asity	Asity, Soy-Soy	o			
183	*Philepitta schlegeli*	Schlegel's Asity	Asity		o		
183	*Neodrepanis coruscans*	Sunbird Asity	Soymanga	o			
183	*Neodrepanis hypoxantha*	Salomonsen's Sunbird Asity	Soymanga	o			
		Larks (Alaudidae)					
185	*Mirafra *hova*	Madagascar Bush Lark	Sorohitra, Boria	o	o	o	o
		Swallows, martins (Hirundinidae)					
186	**Phedina borbonica*	Mascarene Martin	Tsidi-tsidina	o	o	o	o
187	*Riparia paludicola*	Brown-throated Sand Martin	Firifinga	o			
		Drongos (Dicruridae)					
189	*Dicrurus *forficatus*	Crested Drongo	Rendovy, Railovy	o	o	o	o
		Crows (Corvidae)					
190	*Corvus albus*	Pied Crow	Goaka, Gaga Voronkahaka	o	o	o	o

Page ref. in Milon *et al.* (1973)	Scientific name	English name	Malagasy name	Humid east	Sambi-rano	Western savanna	South western subdesert
		Babblers (Timaliidae)					
192	*Naemixis viridis*	Green Jery	Jery	o			
192	**Neomixis striatigula*	Stripe-throated Jery	Jery, Kimimitsy	o			o
194	*Neomixis tenella*	Jery	Jery, Kimimitsy, Zea Zea	o	o	o	o
195	**Hartertula flavoviridis*	Wedge-tailed Jery	Jery	o			
196	**Mystacornis crossleyi*	Crossley's Babbler	Talapeutana	o			
197	**Oxylabes madagascariensis*	White-throated Oxylabes	Foditany	o			
201	**Crossleyia xanthophrys*	Yellow-browed Oxylabus	Foditany	o			
		Bulbuls (Pycnonotidae)					
197	*Phyllastrephus *madagascariensis*	Tetraka	Tetraka	o	o	o	
197	*Phyllastrephus *tenebrosus*	Dusky Tetraka	Tetraka	o			
198	*Phyllastrephus *zosterops*	Short-billed Tetraka	Tetraka	o			
200	*Phyllastrephus *apperti*	Appert's Tetraka	—			o	
201	*Phyllastrephus *cinereiceps*	Gray-crowned Tetraka	Foditany	o			
202	*Hypsipetes madagascariensis*	Madagascar Bulbul	Horovana, Tsakorovana	o	o	o	o
		Cuckoo Shrikes (Campephagidae)					
203	*Coracina *cinerea*	Madagascar Cuckoo Shrike	Kikimavo	o	o	o	o
		Thrushes etc. (Turdidae)					
205	*Copsychus *albospecularis*	Madagascar Magpie Robin	Fitatra, Todia	o	o	o	o
207	*Monticola *sharpei*	Eastern Madagascar Rockthrush	Tsinoly, Androbaka	o			
207	*Monticola *bensoni*	Benson's Rockthrush	—			o	
207	*Monticola *imerina*	subdesert Rockthrush	Vorompotana				o
210	*Saxicola torquata*	Stonechat	Fitatra	o	o	o	o
		Warblers (Sylviidae)					
212	*Cisticola *cherina*	Madagascar Cisticola	Tinty, Tsinstina	o	o	o	o
213	**Nesillas typica*	Tsikirity	Poretika, Aritike	o	o	o	o
214	*Dromaeocercus brunneus*	Brown Emu-tail	Voron-driviky	o			
214	*Dromaeocercus seebohmi*	Gray Emu-tail	Voron-driviky	o			

Page ref. in Milon *et al.* (1973)	Scientific name	English name	Malagasy name	Humid east	Sambi-rano	Western savanna	South western subdesert
216	**Thamnornis chloropetoides*	Thamnornis	Kiritika, Aritike				o
216	*Acrocephalus *newtoni*	Madagascar Swamp Warbler	Boron-bararata	o	o	o	
217	**Randia pseudozosterops*	Rand's Warbler	—	o			
		Flycatchers, newtonias (Muscicapidae)					
218	*Terpsiphone *mutata*	Madagascar Paradise Flycatcher	Tsingitry	o	o	o	o
219	**Pseudobias wardi*	Ward's Flycatcher	—	o			
220	*Newtonia brunneicauda*	Common or Brown Newtonia	Tretretre, Katekateky	o	o	o	o
220	**Newtonia amphichroa*	Dark Newtonia	Tretretre, Katekateky	o			
220	*Newtonia fanovanae*	Fanovana Newtonia	—	o			
222	**Newtonia archboldi*	Archbold's Newtonia	—				o
		Wagtails (Motacillidae)					
222	*Motacilla *flaviventris*	Madagascar Wagtail	Trio-trio	o	o	o	
		Vangas (Vangidae) (ENDEMIC FAMILY)					
224	*Leptopterus viridis*	White-headed Vanga	Tretreky	o	o	o	o
225	*Leptopterus chabert*	Chabert's Vanga	Tsa-tsak, Tsaramaso	o	o	o	o
227	*Cyanolanius madagascarinus*	Blue Vanga	Vorontsara-elatra	o	o	o	o
227	*Schetba rufa*	Rufous Vanga	Siketriala, Karapohovava	o	o	o	
228	*Oriolia bernieri*	Bernier's Vanga	—	o			
229	*Vanga curvirostris*	Hook-billed Vanga	Vanga, Voromareny, Voronbanga	o	o	o	
230	*Xenopirostris xenopirostris*	Lafresnaye's Vanga	Tsilovanga				o
230	*Xenopirostris polleni*	Pollen's Vanga	Kinkimavo	o			
230	*Xenopirostris damii*	Van Dam's Vanga	—			o	
232	*Euryceros prevostii*	Helmet Bird	Siketribe	o			
233	*Calicalicus madagascariensis*	Red-tailed Vanga	Kiboala	o	o	o	
234	*Falculea palliata*	Sicklebill	Voronzaza	o	o	o	o
235	*Hypositta corallirostris*	Coral-billed Nuthatch Vanga	Sokodidy	o			

Page ref. in Milon *et al.* (1973)	Scientific name	English name	Malagasy name	Humid east	Sambi-rano	Western savanna	South western subdesert
236	*Tylas eduardi*	Tylas Vanga	Kinkimavo	o		o	
		Starlings (Sturnidae)					
237	*Saroglossa *aurata*	Madagascar Starling	Vorontainaomby	o	o	o	
238	*Acridotheres tristis*	Indian Myna	Martaina	o		o	
		Sunbirds (Nectariniidae)					
240	*Nectarinia *souimanga*	Souimanga Sunbird	Soy, Sobitiky	o	o	o	o
242	*Nectarinia *notata*	Madagascar Green Sunbird	Soy-mangavola	o	o	o	
		White-eyes (Zosteropidae)					
243	*Zosterops *maderaspatana*	Madagascar White-eye	Ramanjerika, Fotsy-maso	o	o	o	o
		Weavers, fodies (Ploceidae)					
245	*Ploceus *nelicourvi*	Nelicourvi Weaver	Fody-sahy	o	o		
247	*Ploceus *sakalava*	Sakalava Weaver	Fody, Fody-sahy, Draky			o	o
246	*Foudia madagascariensis*	Red Fody	Fody	o	o	o	o
247	*Foudia omissa*	Red Forest Fody	Fody	o			
		Mannikins (Estrildidae)					
248	*Lonchura *nana*	Madagascar Mannikin	Tsikirity, Tsakapia	o	o	o	o

CHAPTER 8

Introduction to the Mammals

RENAUD PAULIAN

Of all the Malagasy animals, the mammals are most obviously unique. Therefore, when biogeographers imagined an Indo-Malayo-Malagasy continent they called it Lemuria, after the extraordinary lemurs of Madagascar.

The Malagasy fauna is equally remarkable for the species which are present and for those which are absent. Let us leave aside the cortege of species which accompany man in his conquest of the earth: rats, mice, shrews, domestic and semi-domestic animals. The rest give us the astonishing picture which follows.

In spite of its ancient origins, Madagascar possesses neither Monotremes (oviparous mammals of the Australian region) nor Marsupials (now relegated to Australia and America, but found in Europe in past evolutionary times).

As for Primates, Madagascar has more prosimians, (the lemurs) and more diverse species of prosimian, than any other region of the world. However, the true monkeys and apes which are so widespread in the tropics of both hemispheres, are totally lacking.

Among the carnivores, only the Viverridae are represented or differentiated — no great cats or wild dogs, no otters or bears.

The rodents, so diverse elsewhere, are here confined to a single subfamily, the *Nesomyinae*.

The ungulates have just one present-day representative, the Malagasy wild boar, of the Suiform suborder, and a subfossil dwarf hippopotamus. The immense group of ruminants which are omnipresent in Africa has never penetrated the island. There you meet neither buffalo, nor antelope, nor giraffe.

In some surprising way a *Plesiorycteropus*, a Pleistocene aardvark, did reach Madagascar. It was a member of the curious group of Tubulidentea, whose sole surviving members are *Oryctoropus*, the aardvarks, which not only live in Africa, but once inhabited Europe and Asia.

There are neither Proboscidians (elephants) nor Hyracoides (hyraxes) nor Perissodactyls (zebras, rhinoceros, etc.) in spite of their frequence in Africa and Asia, neither Edentates nor pangolins nor lagomorphs (hares and rabbits) nor colugos. There is a dugong in the coastal waters, numerous bats, and very special insectivores.

The amazing disequilibrium of this fauna is in part due to the original forest cover of the island. It chiefly results from Madagascar's long isolation during the Secondary Era. Mammals only reached the island sporadically, by accident. They came from East Africa, either on rafts of branches washed out on the flood crests of the great rivers, or else by stages over the islands and shallows which appeared in the Mozambique Channel at epochs of low sea-level.

The few individuals which arrived in Madagascar in this fashion radiated into many species, occupying the empty ecological niches which they found. This evolution in a protected, isolated milieu has given this fauna its peculiar nature and its immense interest.

Fig. 8.1 *Potamochoerus larvatus*, the Madagascar wild boar, is the only wild ungulate. It is very similar to the African species, and was probably introduced by man (A. Jolly)

This introduction to the mammals concludes with notes on some of the lesser studied orders. The three articles which follow will then treat the insectivores, carnivores, and lemurs in more detail.

UNGULATES: THE "LAMBO" OR MALAGASY WILD BOAR

Aside from the dwarf hippopotamus, now extinct (c.f. Chapter 1), a small wild boar (*Potamochoerus larvatus* or lambo in Malagasy) is the only Malagasy representative of the Ungulate order.

The lambo is very close to the African form. It may have reached Madagascar recently, during the Pleistocene with the hippopotamus. It may also have been introduced by man. In East Africa its congeners sometimes live in semi-liberty in the villages.

It is found in the forests of both East and West. Long reddish or yellowish bristles cover its body. One western race has a kind of thick whitish mane, and its black cheeks are framed in white bristles.

The tusks are not so large as those of European wild boars.

Lambo are not particularly fearful of men. They root up the soil in search of roots, insects and worms, and become important crop pests. Villagers hunt them with the aid of specially trained dogs. A big male can reach the weight of 70 kg. The sow gives birth to a litter of up to six young, between October and December.

SIRENIENS: THE SEA-COW OF TROPICAL WATERS

Among the marine mammals, baleen whales and sperm whales were formerly abundant near Madagascar. The people of the Ile Sainte Marie used to hunt whales from their frail dugouts. These animals are now rarely seen.

Similarly the dugong or sea-cow (Lamboharano or Lambondriaka in Malagasy) which lives in the coastal waters of the Indian Ocean, is now very rare because of overhunting, before the species was protected. They are still sometimes seen on the west coast and in the Comoro Islands.

The dugong is a large herbivorous mammal, averaging 2.5 – 3 m in length, with one individual reported at nearly 6 m. Its body is cigar-shaped with forelimbs transformed into flippers lacking any trace of finger nails or claws, and with hindlimbs wholly absent. The upper lip is greatly enlarged and crowned by a fibrous prolongation on the middle of the inner side. It digs in the sea bed with this fibrous structure, to grub up stems of sea-weeds which the animal consumes.

Dugongs are peaceful creatures, which surface to breathe noisily every two or three minutes. They feed during the night. They have only feeble vision, but are sensitive to odours and particularly to sounds. They are supposed to have a gestation period of about a year, and give birth to a single young.

The dugong's pendant breasts, situated between the forelimbs and their genitalia, contributed to the legend of the mermaids — half-woman, half animal, in the legends of the ancient Greeks. Similarly Sakalava legends sing of the loves and the misadventures of men and dugongs.

RODENTS: AN IMPOVERISHED FAUNA

If we leave aside a few imported, cosmopolitan species, the Malagasy rodents fall into only 17 species belonging to 8 endemic genera. They are members of the subfamily Nesomyinae which some authors class with the Cricetinae, others with French fossils of the Tortonian. A Nesomyine rodent has recently been discovered in the Miocene of Kenya.

It is thus an exceptionally poor fauna with little speciation. However, these rodents are very diverse and occupy all the wooded areas of the island, even in the west. Some, like *Brachytarsomys* and *Hypogeomys* are large in size, and important as prey of the endemic carnivores.

CHIROPTERA: BATS WITH INDO-MALAYSIAN LINKS

As one would expect from their powers of flight, bats are better represented and more diverse in Madagascar than the other mammalian groups. Even so there are few species, fewer, for example than in Central Africa. There is one remarkable gap: there are no *Rhinolophus*, a very widespread genus of the Old World.

The bat family is divided into Megachiroptera, large species which generally live in groups and feed on fruit, and Microchiroptera which eat insects or nectar.

In Madagascar the Megachiroptera are represented by the flying fox or fanihy, *Pteropus rufus*. It roosts in groups of trees or in caves, from which it emerges at twilight in massive flights. People relish the fanihy's fatty meat, and hunt it for food. In the South of the Island, the Masikoro fasten the dried burrs of *Harpagophyton* (or *Uncarina*) onto long wooden wands. The burrs are covered with spines that end in recurved hooks. The hooks snag the fur and wings of flying foxes roosting on cave roofs, and thus the bats can be easily captured. In the Ankarana, people smoke out the bats, and knock them down with sticks or branches as they flee the cave.

The biogeographer is just as interested in *Pteropus* as these gourmets! The genus *Pteropus* is, in fact, Asiatic, and does not extend to mainland Africa. The Malagasy form is replaced by similar species on the Mascarenes, the Seychelles, Aldabra, on Anjouan and Moheli in the Comoros, and also on the Isle of Pemba off the Tanzanian coast. This indicates that in spite of their capacity for long-distance flight, these animals in fact are so sedentary that each island has its separate colony. Besides this three other eastern groups of *Pteropus* are respectively present on Anjoua, on Rodriguez, and on Mauritius and Reunion. It is, thus, the type case of an Asiatic group which reaches its extreme western limit on the islands of the Malagasy region. The African Megachiroptera, particularly the genus *Epomorphus,* have not penetrated the region.

Two other genera of fruit-bats are known from Madagascar, *Eidolon,* an afro-arab genus and *Rousettus*, an old-world tropical genus.

The Microchiroptera are more diverse, and fall into three distinct groups.

The genus *Emballonura* inhabits Madagascar as well as the whole Malayo-melanesian zone as far as Samoa. Its distribution resembles *Pteropus.*

A series of African genera have either species which bridge Madagascar and Africa, or species of the same type in Madagascar and Africa. In the latter case the Malagasy species usually seems to have differentiated from the African one, but some times the Malagasy one is more primitive (e.g. *Traenops furcula*). Some 25 species fall into this category.

Finally the *Myzopoda aurita* holds the unique species of an endemic family. It lives in the eastern forest. This is simultaneously a very archaic creature that resembles primitive South American and New Zealand forms, and yet, in some aspects, highly specialized. *Myzopoda* seems to be a relict species. The particular conditions and long isolation of the Malagasy micro-continent have provided this strange bat with a wholly exceptional evolutionary history.

CHAPTER 9

The Insectivores

JOHN F. EISENBERG and EDWIN GOULD

I. INTRODUCTION

The ordinal name Insectivora is somewhat of a misnomer since it implies that members of this taxon feed exclusively upon insects. While it is true that most members of the Insectivora are small and feed upon invertebrates in soil litter, many of which are indeed insects, it is also true that feeding specializations have been evolved so that some members of the group are specialized for feeding on fish, others on crustaceans, and still others on small vertebrates. The Insectivora as an ordinal category is a bit of a waste basket since it lumps together mammals that really share one common set of features: namely the possession of a conservative dentition and skeletal plan. That many families of the Insectivora had rather independent origins is no longer disputed. Indeed, the tenrecoid insectivores may well be split off in an ordinal taxon of their own — the Tenrecimorpha containing two families: The Potomagalidae and the Tenrecidae (Eisenberg, 1981).

The island of Madagascar is presently situated some 425 km (250 miles) off the coast of Mozambique. Contemporary geological evidence indicates that it began its separation from Africa in the late Cretaceous. That the channel separating Madagascar from East Africa was much narrower during the Paleocene is not to be disputed. As a result of the proximity of Madagascar to East Africa, early crossings from East Africa served to inoculate the island continent with mammalian forms which occurred in Africa in the Eocene (Heim de Balsac, 1972). It is clear from an analysis of the modern Madagascan mammal fauna that emigration by various stocks took place at different times. It is not certain when the tenrecoid insectivores first entered Madagascar but fossils that have been identified within the genus Geogale have been discovered in the Miocene of East Africa (Butler and Hopwood, 1957). The tenrecoid insectivores are the most diverse group on the island and their closest living relatives are to be found in Africa, belonging to the family Potomagalidae. The Potomagalidae includes two living genera both adapted for aquatic life. The larger of the two, *Potomagale velox* is one of the largest living insectivorous mammals and is still found in the river systems of West Africa.

The second group of insectivores present on the island of Madagascar are two species of shrew. Extant shrews are classified in the family Soricidae, but if one raises the tenrecoids to an ordinal rank, then the shrews together with the moles would also have their own ordinal ranking as the Soricomorpha. The two Soricoid insectivores on Madagascar were probably transported to the island by early human colonists. They include *Suncus murinus*, the Oriental house shrew and *Suncus madagascarensis* (*S. etruscus*), the pigmy shrew.

Fig. 9.1 *Microgale cowani*. This is one of the smallest species of the genus Microgale and has converged strongly toward a shrew-like lifestyle

II. AN OVERVIEW OF THE TENRECIDAE

The modern tenrecoid insectivores represent an adaptive radiation that has filled several feeding niches on Madagascar. It is convenient to divide the family Tenrecidae into two subfamilies, the Tenrecinae and the Oryzorictinae. The oryzorictine tenrecs possess a soft pelage, long tail, and range in size from the aquatically-adapted *Limnogale* which is about 25 cm (10 in.) long to *Microgale cowani* which is approximately 6.2 cm (2.5 in.) long and weighs less than 12 grams (Fig. 9.1). *Oryzorictes* is semifossorial and burrows into the soil litter seeking out small invertebrates. *Oryzorictes* seems to be confined to areas with alluvial soil and very probably has been adversely effected by intensive rice agriculture and the use of pesticides. Members of the genus *Microgale* have adapted to a variety of niches but in general are surface foragers or litter foragers preying upon small invertebrates. Some species have exceedingly long tails and have a pronounced climbing ability. Of the some 18 species of *Microgale*, most are confined to forested areas. Of the oryzorictines, only the genus *Geogale* has penetrated into the semi-arid scrub lands. The true microgales are in the main confined to more mesic habitats. Species of the genus *Microgale* that have specialized for higher altitudes such as on the central plateau show the capacity to accumulate fat on a seasonal basis and may enter torpor. Lowland coastal form of the genus, however, do not exhibit torpor and appear to remain active year round (see Table 1).

The second subfamily of Tenrecinae includes larger genera ranging from *Tenrec ecaudatus* which may weigh up to one kilogram to *Hemicentetes*, the striped tenrec which may weigh only about 110 grams as an adult (Fig. 9.2). Whereas the oryzorictine tenrecs are shrew-like in appearance and have a respectable tail, the subfamily Tenrecinae is characterized either by a short tail or a virtually tailess condition. In addition, at some time during their life history the tenrecine insectivores possess a spinescent coat. This is especially pronounced in the genera *Setifer* and *Echinops* that have approached in external morphology the form of a hedgehog. *Hemicentetes* possesses barbed detachable quills that are deterrent to small carnivores that may attempt to prey on them. *Tenrec ecaudatus* has spines as a juvenile but tends to lose them as an adult only retaining them as a spinescent crown on the nape (Eisenberg and Gould, 1970).

Species of the subfamily Tenrecinae are all characterized by the capacity to enter torpor during times of reduced primary productivity. This is especially true for the genus *Echinops* which has occupied the xeric western portion of the island and gestivates during the dry season (Gould and Eisenberg, 1966). *Tenrec ecaudatus* is distributed almost over the entire island. This is a highly adaptable species much sought after for food in rural areas. It is completely omnivorous and is characterized by one of the highest litter sizes (32) of a living eutherian mammal.

Setifer and *Echinops*, in their feeding habits and anti-predator mechanisms, have converged strongly with the European hedgehog. *Hemicentetes* is the most specialized genus within the subfamily. It includes two species *semispinosus* and *nigriceps*. *H. semispinosus* is confined to the eastern lowland wet forests whereas *H. nigriceps* is confined to the central, eastern portion of the plateau region. Both species are highly adapted for feeding on earthworms which they probe for in leaf litter with their rather long snout. Their detachable barbed spines and warning coloration serve as an effective antipredator device.

II. SOME CONSERVATIVE TRAITS OF THE TENRECIDAE

In the main, the tenrecs confine their activity to the hours of darkness. *H. semispinosus* is exceptional in having a peak of activity at midday and an activity period after sunset. The eye is rather reduced in size and their reliance on visual signals is minimal. *Tenrec* shows the greatest visual acuity as demonstrated by their immediate body orientation and mouth gape directed toward a quietly approaching object. The tactile, auditory and olfactory senses are highly developed: during foraging, tenrecs rely on tactile, auditory, and olfactory cues to locate prey. They are sensitive to the odours of potential predators and take appropriate antipredator responses when they detect the odours of viverrids such as *Galidia* and *Fossa*. Tenrecs can utilize tongue clicks as a method of echolocation (Gould, 1965). Simple echolocation systems employing tongue clicks, tooth clicking or ultrasonic pulses are widespread in conservative, nocturnal mammals.

The tenrecs generally construct burrows where they nest and rear their young. Nests in a special chamber of the burrow system are constructed of leaves transported by the adult to the burrow from the forest floor. Thermoregulation by tenrecs shows a great deal of variation: all species show some fluctuation during a 24 hour period of their core body temperature. *Microgale talazaci* and *M. dobsoni* have core body temperatures of 31 – 32°C, *Hemicentetes nigriceps* is active with a body temperature at 30°C, while *Setifer setosus* may be active over a range from 30 – 28.7°C body temperature. *Echinops telfairi*, *Setifer setosus*, *Tenrec ecuadatus* and both species of *Hemicentetes* can enter seasonal torpor with a body temperature held about 1°C above the ambient temperature over an ambient range of 19 – 27°C (Eisenberg, 1980; Gould and Eisenberg, 1966).

IV. REPRODUCTION

The testes in the male tenrecine insectivore remain near the kidneys and never undergo descensus. The oryzorictine insectivore male exhibits testicular migration during development and in the adult the testes are lodged in the pelvic cavity. All tenrecs possess a cloaca or common chamber into which the anus and urogenital system discharges.

The gestation period of tenrecs is rather protracted for their body size but post partum maturation is very rapid in the species of the Tenrecinae (see Table 2). Litter size varies widely from species to species reflecting the long-term effects of selection that has produced both "r" and "K" selected forms. A species such as *Tenrec ecaudatus* reproduces rapidly with an early maturation and a large litter size. *Microgale talazaci* matures slowly and has a reduced litter size. The "microgales" represent species with little capacity to respond quickly to environmental changes. A slow reproductive rate renders such species vulnerable to rapid extinction.

V. SPECIES ACCOUNTS

a. *Suncus madagascarensis*

Suncus madagascarensis has been considered a possible subspecies of the continental *Suncus estruscus* (Ellerman and Morrison-Scott, 1951). This small insectivore is widely distributed throughout the island of Madagascar but very seldom seen. As an adult, it rarely exceeds 3 g in weight. It is a specialist in feeding on small insects: despite its diminutive size it can kill and devour a grasshopper slightly larger than itself. It is rarely collected but frequently shows up in owl pellets which have been regurgitated by the owl after feeding on small mammals. By examining owl pellets, one can gain some insight into the island-wide distribution of this cryptic species since the mandible and parts of the skull are readily identifiable from other small mammal remains. Its occurrence (during the dry season) in the extremely dry limestone forest east of Antsalova in the Antsingy Forest attests to its adaptability to extreme environments (Gould and Parcher, unpublished).

b. *Suncus murinus* (Voalavo fotsy)

This species of shrew (the Oriental house shrew) is frequently a commensal of man. Its odour or loud piercing chirps are sometimes detected before seeing this common shrew in the city streets of Tananarive as well as at country villages. Undoubtedly it was transported to Madagascar by early colonists as an unsuspected stowaway on boats and in cargo. The biology of this species is rather well known from the studies in India and throughout the South Pacific (Dryden, 1968; Stine and Dryden, 1977). In moments of distress the mother *Suncus* and young will caravan as each holds the tail or rump of the shrew in front. It is a frequent colonizer of islands as a commensal of man. It is omnivorous in its habits and is a moderately sized insectivore reaching 20 – 30 g in weight. It is likely to be a competitor of some microgales, particularly at the edge of forests where both sometimes occur. Along some streams where *Limnogale* feeds it may compete for nest sites in the bank.

c. *Tenrec ecuadatus* (Tandraka)

Tenrec ecuadatus as implied by the specific name does not possess a tail. It is one of the largest members of the order Insectivora and the largest tenrecoid insectivore on the island of Madagascar. *Tenrec* has at least two distinctive pelage conditions: rather furry specimens occur in the eastern rain forests on the pleateau near Tananarive and at Diego Suarez; moderately to almost entirely spinescent specimens occur south of Moramonga and near Morondava. It has been introduced to several islands in the Indian Ocean including the Comoro Islands, the Seychelles, Reunion and Mauritius. *Tenrec ecuadatus* is an omnivore feeding on a variety of invertebrates and small vertebrates. It is tolerant of a wide range of habitat types existing in xeric as well as mesic regions (Gould and Eisenberg, 1966). *Tenrec* has one of the largest litter sizes of any extant eutherian mammal. 32 embryos have been recorded and one captive specimen gave birth to and reared 31 young. In the wild, mortality of infants is rather high so that when the young become ambulatory and begin to follow the female in her nocturnal foraging, usually only 10 – 15 young are noted. When moving in the forest a mother and her litter seem to move in orderly fashion: 4 – 5 columns follow behind the mother.

The young develop spines shortly after their birth which are arranged in longitudinal rows. The spines in the centre of the back can be vibrated when the animal is startled. The sound is a brief series of pulses ranging from 2 kHz to 20 kHz. At about a month of age, soft underfur begins to dominate and the

Fig. 9.2 *Tenrec ecaudatus*. Adult. Note retention of spines on the crown of the head (C. Wemmer)

spinescence declines. Spines remain on the crown of the head into adulthood (see Fig. 9.3). During the dry season some Malagasy restaurateurs keep 8 or more adult tenrecs in a drum where they remain torpid. From time to time one is removed for cooking. The large masseter muscles are a delicacy.

d. Hemicentetes (Sora; tsora)

The genus Hemicentetes includes two species *H. nigriceps* and *H. semispinosus*. *H. nigriceps* occurs on the eastern escarpment of the central plateau. Its basic colour is black with longitudinal white stripes and a white crown. *H. semispinosus* occurs in the lowland, moist forests of eastern Madagascar. Its basic colour is also black and it possesses longitudinal stripes and a crown colour of yellow. Both species possess barbed detachable quills especially prominent on the crown. Although only 100 – 150 g in weight, they have an active antipredator offensive system and by bucking the head seek to drive the spines into the snout of a potential predator. The spines are barbed and detachable with the exception of a cluster of specialized spines in the centre of the back (Petter and Petter-Rosseaux, 1963). These spines are stridulating quills that vibrate and produce a signal which is in part ultrasonic. Sound pulses from the stridulating organ consist of broad band noise from about 2 – 200 kHz. The sounds are audible to another tenrec at a distance of more than 4 m. To the human ear stridulation resembles the sound produced by dry grass being rubbed and crackled. This sound is employed during various states of arousal and definitely serves as a location sound for young animals when they are following the mother during her foraging (Eisenberg and Gould, 1970). Litter size is small in *H. nigriceps* being about 4 – 5 young. In

Fig. 9.3 Two species of Hemicentetes. Left: *Hemicentetes semispinosus*; Right: *Hemicentetes nigriceps*. Note the soft, dense underfur of the plateau inhabitating. *H. nigriceps*

H. semispinosus, the litters are larger, with up to 10 young. Both species are highly specialized for foraging in leaf litter and actively seek out areas where earthworms are reasonably abundant. The reduced dentition suggests a tendency toward specialization on earthworms and soft-bodied invertebrates as prey items.

e. *Setifer setosus* (Sokina)

The greater hedgehog tenrec, as the name implies, resembles in its external appearance and in its antipredator behaviour, the Eurasian hedgehogs. The animal is of moderate size reaching 300 g in weight and is covered with spines that are nondetachable. During antipredator behaviour, the animal curls into a ball tucking its nose and folding the spinescent skin ventrad to protect the spineless, soft ventrum. The animal is nocturnal with moderate climbing ability. It is a generalized omnivore and is widely distributed in eastern Madagascar. Setifer has adapted to urban life by feeding from garbage cans in Tananarive. Toward the southwest it is replaced by the lesser hedgehog tenrec.

f. *Echinops telfairi* (Tambotriky)

Echinops telfairi, the lesser hedgehog tenrec, is adapted to the xeric southwest. This tenrec sometimes lives in west coast thatched dwellings where they feed on insects. Arboreal, it is a generalized omnivore and insectivore. *Echinops* is a slow but agile climber in dense shrubbery. With their sharp toenails, *Echinops* can hang by a single foot from the rim of a thin flat surface and then regain its position. When disturbed *Echinops* rolls into a ball, even tighter than that of *Setifer*, tucks its head and emits hisses and a crunching noise with its teeth. It undergoes a prolonged torpor during the annual dry season in the southwest. It usually seeks out hibernacula in hollow logs and trees and may pass its time in torpor with several other individuals. The location of hibernating individuals is revealed by sharply striking a hollow standing tree or felled stump; the disturbed animal huffs and grinds its teeth.

g. *Oryzorictes talpoides* (Voalavonarabo)

There are three species of *Oryzoryctes. O. talpodes* is typical. All members of this genus are specialized for semi-fossorial existence. They are all rather small being less than 10 cm in length. The tail is reduced as are the external ears and eyes. The forepaws are modified for digging and possess elongated claws. They have converged in their foraging strategy with the shrew moles of the holarctic. From what little we know of their natural history, they appear to feed on invertebrates in loose soil and are often found in the vicinity of fresh water streams, rice paddies, and large interior lakes (Fig. 9.4).

h. *Microgale talazaci*

There are some 18 named species of the genus Microgale: *Microgale talazaci* is typical. All members of the genus possess a long tail. They range in size from *M. cowani* which is 6 – 7 cm in head and body length to *M. talazaci* which is some 15 cm in head and body length. Some species are specialized for arboreal foraging such ad *M. longicaudata*; others are more strictly terrestrial. All are specialists on invertebrates and especially bark dwelling or forest litter dwelling insects. *M. talazaci* is found in the rain forests of the eastern escarpment. It is semi-arboreal but frequently forages on the ground. It apparently reproduces only once per year and has from two to three altricial young which take approximately six weeks to begin foraging outside of the nest. Trapping evidence suggests that adults may live in a loose pair bond, since opposite sexed adults are often caught near the same location. Very little direct observations have been made of these animals in nature. The counterpart to *M. talazaci* in forests of the central plateau is *Microgale dobsoni*. Both are similar in size but *M. dobsoni* develops a fat storage organ in the tail during the dry season. It feeds on insects amid the leaf litter on moist, friable soils of steep slopes (Fig. 9.5).

It would appear that many species of *Microgale* are highly specialized for discrete micro-environments, and are thus vulnerable to any form of habitat destruction. Also, it would appear that they are vulnerable to competition from introduced *Suncus murinus* and the black rat *Rattus rattus* (Heim de Balsac, 1972). *Microgale* trapping was always more successful in areas more remote from human habitation and disturbance.

i. *Limnogale mergulus* (Voalavondrano)

This tenrec is specialized for an aquatic existence. It has webbed feet and a laterally compressed tail. The pelage is dense. The animal itself is approximately 20 cm long in head and body length. It is found in the vicinity of fast-flowing fresh-water streams and apparently forages for small fish and crustacea. Faeces on rocks reveal undigested exoskeletons of crayfish. The vibrissae are strongly developed on the snout and it is believed that the animal can detect prey under water by vibrations set up by the movements of the prey and perceived through the vibrissae (Malzy, 1965). In its mode of hunting and adaptations, it parallels such aquatic insectivores in Europe as the desmens. Grandidier and Petit (1932) noted the close association of *Limnogale* with the lace plant *Aponogeton fenestralis*. The association is apparent

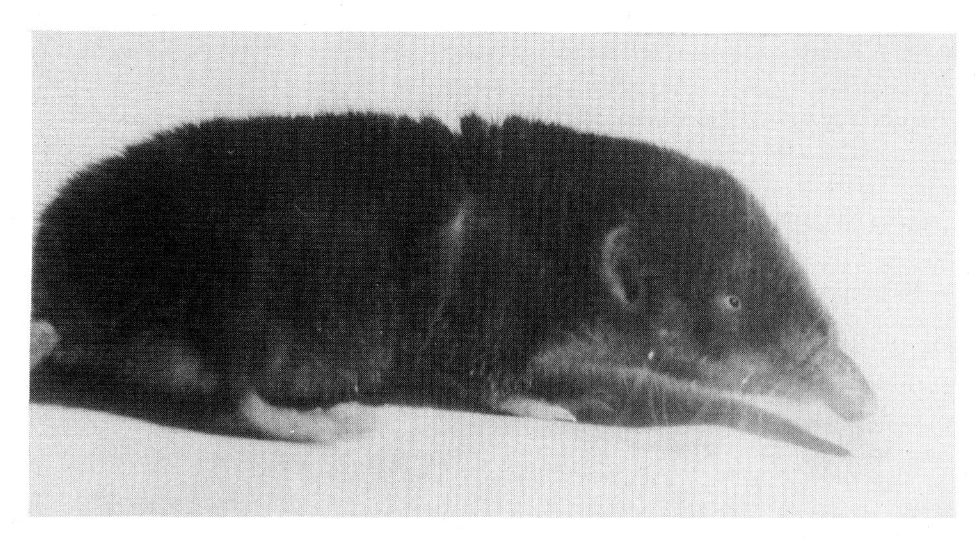

Fig. 9.4 *Oryzorictes talpoides*. Note the elongated snout, reduced eyes and very small external ear. These characteristics suggest adaptations similar to those of the short-tailed shrew in North America

Fig. 9.5 *Microgale dobsoni*. Note the rather large ear and long tail of this plateau-dwelling tenrec. The tail serves as a fat storage organ during periods of food scarcity

at Antsampandrano and Rogez. The rather typical habitat at Rogez (411 m) and the more temperate features of Antsampandrano (1900 m) Forestry Station where introduced trout thrive and ice and snow occur in July and August further emphasize the importance of the lace plant-*Limnogale* association. Thirteen localities for *Aponogeton* extend from Fort Dauphin north to Cap Masoala Maromandia (north of Maroantsetra) and just east of Tulear north to Rapides Mahavavy (south of Majunga) (Kiener 1963). Future zoological studies may reveal a broader distribution of *Limnogale* in association with *Aponogeton*.

VI. CAPTIVE CARE

Echinops, Centetes, Setifer, Hemicentetes, and *Microgale* have been kept in captivity with varying degrees of success. *Echinops* is by far the easiest to maintain and propagate. Diets for captive tenrecs are outlined in Table 3.

During periods of torpor, *Hemicentetes* prefers a damp environment. Tenrecs can hibernate when the temperature ranges from 18 to 22°C. During the breeding season, the animals are maintained at 23 to 26°C (Eisenberg and Gould, 1967). The need for large quantities of earthworms for *Hemicentetes* complicates their care. A quiet nesting chamber, water and adequate space are also critical. *Echinops* can benefit from branches to promote climbing. *Echiniops telfairi* breeds rather easily in captivity. In 1981, 48 specimens were being kept in 11 zoos in the United States.

VII. CONCLUDING REMARKS

The more specialized members of the family Tenrecidae are all vulnerable to human disturbance. The natural predators of the tenrecids include owls, snakes, and the viverrid carnivores. The larger members of the subfamily Tenrecinae have evolved spinescent coats to protect themselves against predation. The smaller oryzoryctine tenrecs rely on cryptic habits to avoid predators.

The tenrecoid insectivores represent a unique radiation of mammals possessing a conservative body plan. Clearly they have been isolated on Madagascar for considerable periods of geological time. In their adaptive radiation, they have occupied niches in a unique manner not replicated by insectivores on the contiguous continental land masses. Although there are parallels in adaptation with insectivores from Africa, Europe and Asia, such forms as *Hemicentetes* and the semi-arboreal Microgales are without parallel.

Several of the Insectivora seem relatively immune to human disturbances. *Suncus murinus* is a commensal. *Echinops* is occasionally a commensal in some huts of small villages. Setifer feeds on urban garbage. *Hemicentetes nigriceps* will nest and sleep during the day in eucalyptus groves and feed on earthworms in nearby grassy rice paddies. However, both insectivorous and earthworm-eating Insectivora are vulnerable to the use of pesticides. Clearing of forests for lumber and agriculture is the major threat to most species of *Microgale*. Silting of rivers and other water pollution would threaten Limnogale populations. Stream pollution, use of insecticides, continued habitat destruction and the introduction of commensals of man such as the oriental house shrew and the black rat will no doubt continue to have an adverse impact on the survival of this unique, relictual radiation of mammals.

REFERENCES

Butler, P.M. and Hopwood, A.T. (1957) Insectivora and Chiroptera from Miocene Rocks of Kenya Colony. *Fossil Mammals of Africa*. British Museum of Natural History, 13, 1–35.

Dryden, G.L. (1968) Growth and development of *Suncus murinus* in captivity on Guam. *Journal of Mammalogy* 49 (1), 51–63.

Eisenberg, J.F. (1980) Biological strategies of living conservative mammals in *Comparative physiology: Primitive mammals*. Eds. K. Schmidt-Nielsen, L. Bolis, and C. Richard Taylor. p.13–30. Cambridge, Cambridge University Press.

Eisenberg, J.F. (1981) *The Mammalian Radiations*. The University of Chicago Press, Chicago.

Eisenberg, J.F. and Gould, E. (1967) The maintenance of tenrecoid insectivores in captivity. *Intern. Zoo Yearbook* 7, 194–96.

Eisenberg, J.F. and Gould, E. (1970) The tenrecs. A study in mammalian behavior and evolution. *Smithsonian Contrib. Zool.* 27, 1–137.

Ellerman, J.R. and Morrison-Scott, T.C.S. (1951) *Checklist of Palearctic and Indian Mammals 1758–1946*. London, British Museum.

Gould, E. (1965) Evidence for echolocation in the Tenrecidae of Madagascar. *Proc. Amer. Phil. Soc.* 109, 352–60.

Gould, E. and Eisenberg, J.F. (1966) Notes on the biology of the Tenrecidae. *Journal of Mammalogy* 47(4), 660–86.

Grandidier, G. and Petit, G. (1932) *Zoologie De Madagascar*. Societe D'Editions, Paris.

Heim de Balsac, H. (1972) Insectivores in *Biogeography and Ecology in Madagascar*. Eds. R. Battistini and G. Richard-Vindard. p.629–60. W. Junk B.V. Publishers, The Hague.
Keiner, A. (1963) Poissons, peche et pisciculture a Madagascar in *Centre Technique Forestier Tropical*. 244p. Seine, France.
Malzy, P. (1965) Un Mammifere aquatique de Madagascar: Le *Limnogale Mammalia*, 29, 400–12.
Petter, J.J. and Petter-Rousseaux, A. (1963) Notes biologiques sur les Cententinae. *La Terre et la Vie* 17(1), 66–80.
Stine, C.J. and Dryden, G.L. (1977) Lip-licking behavior in captive musk shrews, *Suncus murinus*. Behaviour 62(3–4), 298–314.

TABLE 1. Morphological correlations among the terrestrial Oryzorictinae. From Eisenberg and Gould, 1970.

	Head and body length (mm)			
	>40 ⩽60	>60 ⩽85	>90 ⩽115	>115 ⩽135
Fossorial or semifossorial				
T/HB ⩽.50		*Geogale aurita*		*Oryzoryctes hova*
HF/HB ⩽.20		*Microgale brevicaudata*		*O. talpoides*
Surface foragers, moderate climbing ability				
T/HB ⩾.60 ⩽1.00	*M. (Paramicrogale)*	*M. crassipes*	*M. (Nesogale) dobsoni*	
HF/HB >.13 ⩽.20	*occidentalis*	*M. cowani nigriceps*	*M. (Leptogale) gracilis*	
			M. thomasi	
HF/HB >.20 <.24		*M. cowani*		
HF/HB >.25 <.29		*M. longirostris*		
Surface foragers and climbers				
T/HB >1.00 ⩽1.60				
HF/HB >.13 ⩽.20				
HF/HB >.20 <.24		*M. melanorrhachis*		*M. (N.) talazaci*
HF/HB ⩾.24 ⩽.27	*M. pusilla*	*M. drouhardi*		
Climbers and springers				
T/HB >1.60 <2.6	*M. prolixicaudata*	*M. majori*		
HF/HB >.265 <.30	*M. longicaudata*	*M. sorella*		
		M. principula		

T/HB = Ratio of tail to head and body length
HF/HB = Ratio of hindfoot to head and body length

TABLE 3. *Suggested foods for captive tenrecs*

	Earthworms	Mealworms & crickets	Fruits	Raw, chopped horsemeat, condensed milk, baby cereal mix, and vitamins
Tenrec	X	X	X	X
Setifer	X	X	X	X
Echinops		X		X
Hemicentetes	essential			+ egg noodles
Microgale		X		X

TABLE 2. Reproductive data for the Tenrecidae

Species	Gestation (days)	Age of eye opening (days)	Weight of neonate (g)	Range or mean litter size	Average interbirth interval (months)	Average litters /year	Age at first mating (days)	Duration of lactation (days)	References
Migrogale talazaci	58 – 63	18	3.6	1 – 3	—	—	—	30	Eisenberg 1975
M. dobsoni	62 – 64	—	4	1 – 2	—	—	—	—	Eisenberg 1975
Hemicentetes semispinosus	57 – 63	8 – 10	6.5	2 – 11	3.7	2	35 – 40 +*	20	Eisenberg 1975
H. nigriceps	55	8	6.3	2 – 4	3.7	2	30 – 35	20	Eisenberg and Gould 1970
Setifer setosus	65 – 69	13	24.7	1 – 5	—	1	6 mo	—	Eisenberg and Gould 1970
Echinops telfairi	62 – 65	(7 – 9)	6	1 – 10(5.8)	—	—	6 mo	30	Eisenberg and Gould 1970
Tenrec ecaudatus	57 – 63	9 – 14	25	1 – 32	3.4	1 – 2	6 mo	29	Eisenberg and Gould 1970

CHAPTER 10

The Carnivores

ROLAND ALBIGNAC

There are seven indigenous species of Malagasy carnivores, and three more species introduced by man. Only one of the foreign species is a wild animal, the palm civet *Viverricula rasse*. It lives in the savannah, mainly near villages, and has never penetrated the forest. The other two, the cat and dog, are domestic though with feral individuals. The forests, with their myriad endemic plants and animals, remain the domain of native carnivores.

They are relatively few in number compared to the diversity of African and Asian carnivores. They are also few compared to the extensive radiations of other groups such as lemurs and insectivores.

The seven species are classed in three endemic subfamilies: Fossinae, Galadiinae and Cryptoproctinae, all belonging to the nearly worldwide family Viverridae. They share some close anatomical characteristics, or ones which are clearly derived from each other, which suggest a monophyletic origin for the group. Their common ancestor probably crossed the Mozambique Channel from Africa.

Classification of endemic Malagasy carnivores

Family Viverridae
- Subfamily Fossinae
  - Genus *Fossa* G.E. Gray 1864
    - *F. fossana* P.L.S. Muller 1776
  - Genus Eupleres Doyere 1835
    - *E. goudotii goudotii* Doyere 1835
    - *E. major* Lavauden 1921
- Subfamily Galidiinae
  - Genus *Galidia* I. Geoffroy Saint-Hilaire 1837
    - *G. elegans elegans* I. Geoffroy 1837
    - *G. e. dambrensis* G.H.H. Tate, A.L. Rand, 1941
    - *G. e. occidentalis* R. Albignac 1971
  - Genus *Salanoia* G. e. Gray 1941
    - *S. concolor* I. Geoffroy 1839
  - Genus *Mungotictis* R.I. Pocock 1915
    - *M. decemlineata decemlineata* A. Grandidier, 1867
    - *M. d. lineata* Pocock 1915
  - Genus *Galidictis* I. Geoffroy Saint-Hilaire 1839
    - *G. fasciata fasciata* Gmelin 1788
    - *G. f. striata* E. Geoffroy 1826

Subfamily Cryptoproctinae
Genus *Cryptoprocta* Bennett 1833
C. ferox Bennett 1833

All these indigenous carnivores are distinctly forest animals, except the fosa (*Cryptoprocta ferox*), whose Malagasy name is pronounced "foosh". It dens in the forest but can venture far across open country in search of prey.

Other characteristics beside forest habitat distinguish Malagasy from African viverrids. Most Malagasy carnivores are monogamous, though with some variation. Only the fosa differs sharply, by having solitary habits. Malagasy carnivores also have unusual reproduction for viverrides, bearing only one young per year, again except the fosa which bears a litter of 2–4.

THE FOSSINAE

The first subfamily the Fossinae, has two fairly closely related species, *Fossa fossana* and *Eupleres goudoti*, the fanaloka. (The fosa *Cryptoprocta ferox*, should not be confused with *Fossa fossana*, though Gray presumably did so in 1864 when he gave *Fossa* its latin name.)

These two species are nocturnal and purely terrestrial. They live mainly in humid habitats, *Fossa* along watercourses in the eastern forest, and *Eupleres* in marshes with *Raphia* and *Pandanus* trees in the east and the Sambirano.

The two are much alike in morphology, with heavy bodies and fairly short, cylindrical tails. The paws are thin with medium sized claws in *Fossa*, but large, powerful claws in *Eupleres*. The muzzle is long and thin, with fairly undifferentiated dentition in *Fossa* and reduced dentition in *Eupleres*.

They are both capable of laying down reserves of fat as winter approaches, on the body and especially on the tail. These reserves can reach a quarter of their body weight. They do not hibernate. However, the fat helps them over a severe food deficit from June to August.

Fossa fossana has dense fur. It is light brown, spotted with darker brown, the spots more or less in lines on the body, flanks and tail. Its teeth are shearing type. The carnassial premolar is clearly differentiated but the canines little developed. The dental formula is

$$I\ \frac{3}{3}\quad C\ \frac{1}{1}\quad Pm\ \frac{4}{4}\quad M\ \frac{2}{2}$$

Fossa is fairly rare and always secretive. During the day it shelters in family groups in hollow logs or in rock clefts. Home ranges are large — several hundred hectares — and seem very stable over several years.

The *Fossa* has a digitigrade gait, which helps it capture small prey after a rapid chase. Its jaws open very widely, which helps it seize the prey more securely.

Fossa catches rodents in this fashion as well as frogs and even eels which it seizes without apparent difficulty as they wriggle through shallow water a few centimetres in depth. It also eats earthworms, insects, crustaceans, and sometimes birds. It seizes and kills prey with the jaws alone, and only uses its forepaws to hold dead prey on the ground.

This species has few perceptible means of vocal communication. They can be summed up as muffled growls, given rarely in the course of agonistic encounters. There is also a special cry given by parents and young during the 4–5 months after birth.

Olfactory signals, on the other hand, are much commoner, with visible marking behaviour. There

Fig. 10.1 *Fossa fossana* (R. Albignac)

Fig. 10.2 *F. fossana*, with young about 2 weeks old. Its eyes are open and it is densely furred at birth (R. Albignac)

is a glandular ring round the anus with large groups of sebaceous cells. The animal rubs this gland on various types of surface during territorial marking.

Reproduction is seasonal. Courtship preliminary to mating lasts 15 – 20 minutes. Gestation is about 3 months (82 – 89 days in captivity). Births generally occur during November and December, the austral summer.

The single young, unlike that of many other carnivores, is already well developed at birth. Its eyes are open and it is fully furred, though somewhat darker in colour than the parents. It weighs 60 – 70 g.

The female excludes the male from the den at parturition. She raises the young alone during the first month. She carries the baby by picking it up in her jaws by the scruff of its neck. The little one then rolls up in a ball, with limbs folded to its body.

The young develops relatively slowly. It walks at 3 days and begins to eat meat a month after birth, but is not weaned until 2.5 months of age. It can seize small prey like frogs and insects from the fourth month onward, but does not catch the same prey as the adults until a year of age. It is probably at this age that the young leaves its parents in the wild.

Eupleres goudotii is a more or less dark grey brown, with lighter stomach, and a somewhat orange area on the thigh. It has large claws on the forepaws which it uses to dig when searching for worms and arthropods. The claws also aid in defence. They are not retractile; to keep from wearing them down the animal holds them up as it walks, which gives it a heavy, ungainly gait. The dentition is extraordinarily small for a carnivore. The teeth are tiny and conical or flattened.

Eupleres may be fairly abundant locally but on the whole is rare or even very rare. People frequently

Fig. 10.3 *Eupleres goudoti*, the fanaloka. Its chief food is earthworms (R. Albignac)

Fig. 10.4 *E. goudoti* with young about 1 month old. Like *Fossa,* the young are well developed at birth (R. Albignac)

set traps for it, since its meat is much appreciated, which both helps cause and helps reveal its rarity.

People say that *Eupleres* is able to dig burrows for daytime refuge, or even for hibernating through the June-August winter months. In fact, it seems incapable of digging burrows. In captivity it always lies down on the ground, wherever low-level vegetation is thickest.

Its chief food is earthworms, which are also the preferred food in semi-captivity. The long claws and muzzle are very efficient in catching earthworms, and the reduced dentition is very efficient in cutting such prey into small pieces to swallow. *Eupleres* also eats insects and frogs, and in captivity, accepts meat cut into small strips.

It makes even less frequent vocalizations than *Fossa*. It gives only one defensive call, a kind of spitting. After parturition as in the *Fossa*, the mother has a special call and the young a kind of mewing. These calls are essentially the same in the two species and permit mother-young coordination in the course of nocturnal movements.

Olfactory signals are very like *Fossa*'s: the same marking behaviours and glandular zones. Home range is, likewise, very large.

Birth occurs in summer. The young is well developed: fully furred and much darker than the adult, so dark brown it is nearly black. The eyes are open at birth. Mother-young behaviour is like *Fossa's*, and it is possible that the young leaves the parents at about a year old.

THE GALIDIINAE

The second subfamily, Galidiinae, has four species rather like small mongooses, both in morphology and in mongoose-like ways of life.

They are only about 50 cm long, of which more than 20 cm is tail. Weight is about 600 – 1000 g. They are all diurnal except the nocturnal *Galidictis*. They are both arboreal and terrestrial, with glabrous soles whose large pads give them assurance and agility in the trees. The muzzle is shorter than in the Fossinae, the carnassial is larger, associated with reduction or loss of some of the other teeth. The dental formula is

$$I \frac{3}{3} \quad C \frac{1}{1} \quad Pm \frac{3}{3} \quad M \frac{2}{2}$$

Galidia (vontsika or kokia) has several colour forms, often linked to chromosomal variations. The general colour is a handsome russet brown, lighter or darker in the different subspecies. The tail, the same colour as the body, always has five or six darker brown rings.

In the subspecies *G. elegans elegans*, which lives in the evergreen forests of the east, the body is a

Fig. 10.5 *Galidia elegans* is convergent with mongooses, in body form and omnivorous diet including snakes (R. Albignac)

Fig. 10.6 *G. elegans* with 10 day old young. The newborn's eyes do not open until the 6 – 8th day, and it cannot walk until about the 12th day (R. Albignac)

lovely deep red brown, and the ventral surface is darker brown speckled with grey hairs. The extremities of the limbs are black.

In the subspecies *G. elegans dambrensis*, localized in the north the body is light russet, and the stomach, paws and flanks are about the same colour as the body.

Finally, in *G. elegans occidentalis*, found in all the karst formations of the west, the body is likewise light russet but the ventral surface, paws and flanks are pure black.

All these animals are relatively common, and it is not unusual to see them in the forest.

Galidia usually shelters in burrows which it digs itself. It can also occupy hollow logs, sometimes at several metres height, or even in cracks between fallen rocks.

The animal usually travels on the ground but it often climbs large trees in search of arboreal prey (either small vertebrates or arthropods) or sometimes with no apparent reason. If it has to climb a trunk more than 3 cm in diameter it goes up like an inchworm, head upwards. If the trunk is smaller it more or less walks up, with a forefoot moving at the same time as the opposite hindfoot. When descending its head is downwards and it moves inchworm-fashion whatever the size of support. In the course of arboreal locomotion, the first and especially the second finger play an important role. The first finger is set almost in opposition to the others, while the claw of the second hooks strongly into the support.

On the ground, *Galidia* can make sudden about-turns in the course of walking or running. It also enters water to seize frogs and fish.

Galidia breaks eggs and snails, mongoose fashion. It seizes the egg in its forefeet, lies down on one side, brings up its hindfeet and passes the egg back to them, then, with a brusque kick, smashes the egg on a trunk or wall. After this it licks up the contents of the broken shell.

Galidia commonly eats rodents which it actively chases, either digging out their burrows or systematically prospecting through trees. The diet, however, is extremely varied: insects, worms, snails, eggs and even reptiles and small lemurs.

Vocal communication is varied in comparison to the other Malagasy carnivores. While moving the animal gives a little whistle like a bird, which seems to be a contact call. There is a variety of meows. Short, muffled, repeated calls are given when capturing prey. Calls of aggression and defence range from a low growl to a loud shrill scream which can last 4 – 5 minutes.

Olfactory signals and marking are also common. Marking behaviour is identical to that of the Fossinae. There are large anal glands and even a rudimentary perineal gland which only exists in the Galidiinae.

Galidia lives in small family groups of 3 or 4 individuals or more. When moving, the female usually leads and the male follows her or brings up the rear if there are young. The animals of one group always share the same den. During the day they spend long periods playing.

The mating season is limited to the end of austral winter and spring, between July and November.

Gestation is slightly shorter than 3 months (74 – 90 days in captivity). Births occur between August and April, but peak between November and January.

The newborn has fur which is about the same colour as its parents. The eyes, however, are closed and do not open until the 6th to 8th day. The newborn cannot walk, and weigh about 50 g.

As in *Fossa*, the female excludes the male from the usual den immediately after parturition, and raises the young alone for the first weeks. At first the male is not allowed nearer than 2 metres to the young. This distance gradually lessens, and the female finally lets the male approach his offspring after a month and a half. During the first month the young may also be carried by the neck between his mother's teeth.

The young develops more slowly than in the Fossinae. It only begins to walk at 12 days, and begins to be interested in meat at one month, but weaning is not complete until 2 – 2.5 months.

At the age of three months the juvenile easily seizes grasshoppers and sometimes little frogs. It plays a great deal with its parents; these games are essentially mock attack and defence.

At 10 months the juvenile can fish, but it only begins to catch rodents around 14 months. Around this time it begins to leave its parents, and has left at the latest at 2 years.

Mungotictis decemlineata (boky-boky) also has colour variants in different localities. It is thus reasonable to separate the species into two subspecies.

The muzzle is relatively short in *Mungotictis*, and the mobile ears are obvious. The tail is almost as long as the body. It is not ringed but ends in long hair like a paint-brush. The animal lifts its tail and bristles the hair in threat behaviour. The fur is dense, mainly grey, sometimes tinged with beige; the back and sides have more or less sharply clear dark brown streaks. The tail is uniformly whitish or grey-beige.

Mungotictis lives only in deciduous woodlands growing on sand in the west and southwest of Madagascar. The Tsiribihina river is its southern limit. In the Morondava region they seem to be relatively common animals, frequently seen in the course of a day in the forests. Further north they are not so abundant.

This animal shelters in burrows during the dry season (June to October). The burrows are shallow, and dug by all individuals of a group together. They install themselves at the bottom of a simple oval chamber with only one exit.

Fig. 10.7 *Mungotictus decemlineata* raises its tail, with bristling hair, in defensive threat (R. Albignac)

Fig. 10.8 *M. decemlineata* holds a snailshell with its paws, and breaks into it with its teeth (R. Albignac)

Fig. 10.9 *M. decemlineata* noses through the litter in search of insect larvae. Note radio-tracking collar, and tail clipped for identification (R. Albignac)

During the rainy season (November to May) they sleep in a hollow tree, sometimes 10 m above ground level.

Mungotictis feeds on little rodents and lemurs, and also an assortment of other prey, including worms, insects, snails and reptiles. The diet is thus varied, but with a larger proportion of insects than in *Galidia*, especially during the dry season.

A 14 month field study has shown that insectivorous tendency is much more important than previously supposed. *Mungotictis* essentially eats insect larvae which it gathers from the soil and from decaying wood. These larvae are its mainstay in the cold dry season. In the hot wet season when many small vertebrates become active, its diet becomes more varied. *Mungotictis*, like other vertebrates of the western forests, thus has evolved dietary adaptations to seasonal selective pressures.

Its social organization also changes during the yearly cycle. During the rainy season small social groups of 10 – 12 individuals occupied a home range of about 150 ha. During the dry season, they separated into 2 or 3 subgroups which maintained smaller ranges, though within the same general area. Some of these were even males who associated with juveniles. At the beginning of the next rainy season the subgroups fused again, to form the same social groups as before. The adult population must be very stable and so must the range of each social group. Five years after the first 14-month study the area was retrapped — there were no changes of individuals within the known social groups.

Salanoia lives on the east coast of Madagascar. It resembles *Galidia* in morphology, but seems to have a diet like *Mungotictis*, which would avoid feeding competition.

Finally the last of the Galidinae, *Galidictis*, lives in the eastern rainforest. Its biology is at present totally unknown.

CRYPTOPROCTINAE

The last endemic subfamily, the Cryptoproctinae, has only one species, the fosa (*Cryptoprocta ferox*). It is the largest of the natural Malagasy predators, reaching 1.5 m long including tail, and a weight of 7 – 12 kg.

The fosa occurs throughout the island, except in a small area of the High Plateaux where human pressures are probably too great. In spite of this wide distribution there are no local variants. People occasionally report melanistic forms, but there is no specimen to confirm these stories.

It is low slung, on heavy paws. The claws are retractile, the gait plantigrade or semi-digitigrade. The tail is almost as long as the body. The muzzle is short and square, the forehead large, and the ears large and clearly set off at their bases.

The general colour is red brown, sometimes shading toward grey on the back, without stripes or spots. The fur is dense, short and coarse. The tail is well muscled, and does not accumulate fat reserves: it plays an important role in arboreal locomotion.

The dentition is the cutting type, with well developed canines and carnassials. The upper molars are almost wholly reduced. Both dental formula and general appearance of the teeth recall the Felidae:

$$\mathrm{I}\ \frac{3}{3}\quad \mathrm{C}\ \frac{1}{1}\quad \mathrm{Pm}\ \frac{3}{3}\quad \mathrm{M}\ \frac{2}{2}$$

Activity is essentially crepuscular and nocturnal, but fosa are not infrequently seen by day, particularly toward the end of the afternoon. They live in all the wooded areas of the island, and in part on the savannah, where they venture out by night in search of prey.

The fosa occupies various kinds of shelter. It may sleep in the forks of a large tree, and seems to

Fig. 10.10 *Cryptoprocta ferox*, the fosa. This cat-like viverrid is at home in the trees or on the ground. It descends trunks head-first, braced by hind legs and tail. The claws, like cats' are retractile (R. Albignac)

prefer such lofty refuges. All the same, it can excavate in the earth, and one litter of fosa young was found inside an old termite hill which had been hollowed out as a nest. You may also find fosas in caves.

The fosa is both terrestrial and arboreal. Paws and tail help it climb with ease, and even leap from branch to branch. The soles of the paws are glabrous with raised palmar and plantar pads, as in the Galidiinae. The tail, as long as the body, serves for balance and even as a brace during climbing, or to help break a vertical descent by more or less wrapping round the support.

The fosa can climb trunks at least as large as 80 cm in diameter. It goes up like an inchworm: the forelimbs are spread wide and the hindlimbs folded under the body, propelling the animal forward. To descend the opposite happens: the forelimbs are flexed and the hindlegs spread wide to break the descent. When climbing along lianas with a diameter of 3 – 7 cm, the animal keeps three points of contact with the support. It alternately moves one forefoot forward, then one hindfoot.

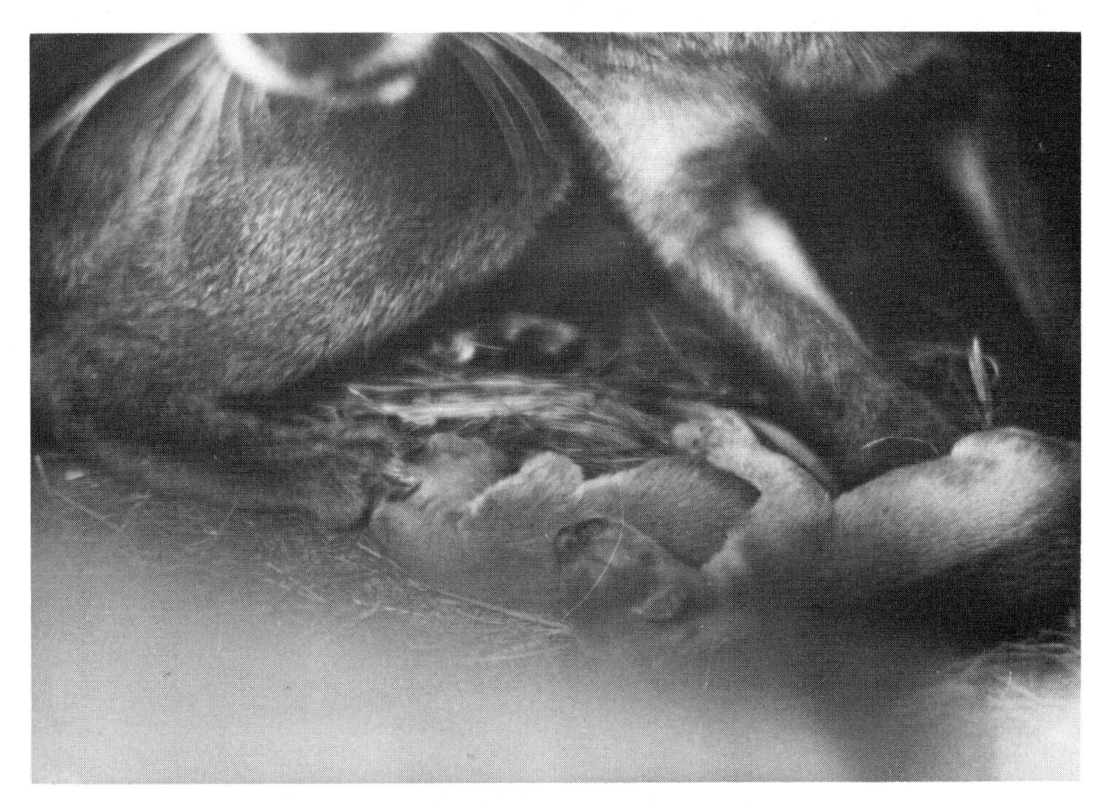

Fig. 10.11 *C. ferox* 5 days old. Unlike other Malagasy carnivores they are born in litters of 2 – 4, and develop slowly. Their eyes open at about 15 days; they walk at about a month (R. Albignac)

These different means of arboreal locomotion are only used over short distances. For travel further than about 50 m, fosa revert to terrestrial locomotion.

Fosa often use their front paws to immobilize prey, both to capture the prey and to hold it before the kill, which is by biting on the back of the head. The other Malagasy carnivores never use this method, but seize prey directly in their jaws.

Fosa mainly eat small mammals and birds, which they readily accept in captivity. Their terrestrial – arboreal locomotion probably lets them capture both lemurs and roosting birds.

Vocal communication is little differentiated. There are saccadic growls of threat, which grade into prolonged meows. The young often make an audible purring as they begin to suckle and during social games.

Olfactory signals and marking are clearly visible. Fosa rub the anal region on flat surfaces, including the ground. A second type of marking is used for vertical surfaces, especially tree-trunks. Animals (particularly males) rub the genital region and preputial gland on the trunk.

The social behaviour of fosa is little known. It seems to be a solitary animal. However during the mating season one can find 3 or 4 males around a female in oestrus.

Reproduction is seasonal, as in other Malagasy viverrids, and lasts from September to November. The preludes to courtship include long, shrill meowing (again like cats). The first coitus is long and may last an hour.

Births occur from November to January. Unlike other endemic Malagasy carnivores there are 2 – 4 young per litter, although, as usual, there is only one litter per year.

At birth the young are densely furred, but they are pale grey, almost white. Their eyes are closed and do not begin to open until the 15th day. The newborn cannot yet walk and weigh about 100 g. The mother probably raises the young alone.

The young develop slowly in comparison with the other Malagasy carnivores. Weaning takes place at 4.5 months, and adult body length is only attained at 2 years. It takes 3 or 4 years to reach full adult weight.

After birth, the young remain constantly in their nest or shelter. You must wait a month to see them start to take an interest in objects around them. It is about then that they start to play and begin to walk more or less efficiently.

At two months they begin to climb along branches, and to make little jumps. At 3.5 months they can make bounds 2 m long and leap without hesitation from one branch to another. They continue developing slowly and are weaned at about 4 months.

Adulthood arrives even more slowly, at perhaps 3 years, but the young leave their mother much earlier, before the birth of the next litter.

Fig. 10.12 *C. ferox* with month-old young (R. Albignac)

Fig. 10.13 *C. ferox*, 2 months old (R. Albignac)

CHAPTER 11

The Lemurs

ALISON JOLLY, ROLAND ALBIGNAC and JEAN-JACQUES PETTER

Origins and Evolution

The lemurs of Madagascar are a separate experiment in primate evolution. As Thomas Huxley wrote: "there is one true structural break in the series of forms of simian brains; this hiatus does not lie between Man and the man-like apes, but between the lower and the lowest simians, in other words between the old and new world apes and monkeys, and the Lemurs". Only on Madagascar have these primitive primates evolved social life in permanent troops, and diurnal, fruit-eating habits, in parallel with the monkeys and apes of the larger continents. They give us an outside view into the evolutionary potential and the ecological forces which shaped our own ancestors.

The earliest lemur-like fossils come from the Eocene of Europe and North America. These fossils, the family Adapidae, were the "first primates of modern aspect" (Simmons, 1972). The Adapidae had even more elongated snouts than present-day lemurs, and even smaller brain size for their body weight. They probably relied on smell far more than vision. Lemurs today also rely on scent in their daily lives, and their brains contain less visual cortex than in the true monkeys of the old or new world. There are differences between lemurs and monkeys in the bones which compose the middle ear, and differences in the hand — in lemurs the fourth, "ring" finger is longest, in monkeys, apes, and men the third. Lemurs have a tooth – comb. The lower canines and incisors lie flat, facing forward. They are used to chisel gum from trees and to groom other lemurs' soft fur. The important difference, though, is the small brain. In the continents where they competed with monkeys, it seems monkeys simply outsmarted them.

Madagascar had already drifted away from Africa when the Adapids were in their heyday during the Eocene. They may have reached the island by rafting on logs or mats of floating vegetation, clinging with all four hands to the wave-washed twigs. Island bridges may have appeared during periods of lowered sea-level. It is not certain whether there was one invasion or several. The smallest lemurs resemble small African bushbabies so closely they may well share recent ancestry — or they might just both have stayed in an ancient primate niche. Ian Tattersall even argues that modern *Hapalemur* and *Lepilemur* are so close to fossil Adapids that they had diverged before ever reaching Madagascar.

The Mozambique channel slowly widened. By the time true monkeys evolved in the Oligocene, some 30 million years ago, they could not cross to Madagascar. On the larger continents, prosimians including bushbabies, pottos, and the slow and slender lorises, remained nocturnal, insectivorous, and foraged alone. Their social life means sharing a day-nest, answering long-distance calls, or noting a trail of smelly scent-marks. Only on Madagascar were lemurs free to occupy niches of monkeys and even ungulate herbivores if they could.

There are three to five families of lemurs, depending on your preference for splitting or lumping. There are 12 – 13 living and 6 extinct genera, about 26 living species and 12 extinct. Several of the species are further subdivided into rings of subspecies round the island, with breaks commonly at the major rivers.

This is remarkable diversity, especially since the extinct forms are only subfossil. They survived at least to a thousand years ago, on two radiocarbon dates (Tattersall, 1973). They coexisted with the living lemurs and with the earliest human settlers. Their bones are found in the same fossil layers as the pots that cooked them. They still shamble through Malagasy legends, and may even have remained until European settlement. Etienne de Flacourt, in the 1650's, reported the existence of the "Tratratratra, large as a two year-old calf, with a round head and a man's face, front and hind feet like a monkey, frizzy fur, a short tail, and ears like those of a man".

Evolution continues today. The ring of *Lemur* species and subspecies shows a variety of chromosome differences which suggest the various populations are actively diverging. In fact, the ring of wet and dry forest of Madagascar acts almost like an archipelago of islands, in which evolution runs more quickly than either fully divided islands or continuous mainland. This is one clue to Madagascar's riches of lemurs as well as all its other burgeoning forms of life.

If you travel to the eastern rainforest at the level of Perinet where low-lying coastal species meet those of higher altitudes you might find as many as 10 kinds of lemurs in a single forest. The large, leaf-eating indriids would include indri, largest of living lemurs, coexisting with diademed sifaka, then two species of true *Lemur*, and variegated lemur and bamboo-chewing hapalemur, all filling different diurnal niches. Five more nocturnal ones would emerge at dusk to pursue their varied lives. In the dry woodlands of the west or the spiny southern desert you could see almost wholly different species — other representatives of the Indriid, Lemurid and nocturnal families. Only if you travelled round the whole circumference of the island looking in every forest, could you find the whole dazzling array. As the forests are lost, the species of lemurs die with them.

Cheirogaleidae: the little lemurs

The lesser mouselemur, *Microcebus murinus*, is smallest of living primates, weighing a mere 45 – 90 g. It lives throughout the forested areas of Madagascar, with a dark rufus race in the gloom of the eastern rainforest and a greyer form in the moonlit west and south. These two may in fact be separate species. It, or they, prefers to run on the twigs of bushes and small lianas, in both primary and secondary forest and on the edges of paths and clearings. It resembles the earliest of primates, and may become the last of Malagasy lemurs to survive in the wild.

Lesser mouselemurs eat a high energy diet of insects and ripe fruit. They also conserve energy as their body temperature fluctuates with ambient temperature, and they can enter a somnolent state in the dry season when their temperature drops, though they still wake once a day to feed. They bear twin young, and may have two litters a year, kept in a nest or tree hollow. The infants are born fully furred, but with closed eyes, and the mother carries them in her mouth if she moves them at all. They move about awkwardly and begin to eat solid food at about three weeks of age.

The mouselemur social system seems to be a harem by location — probably the original primate society. The home range of one large male overlaps the smaller ranges of several females. Transient, or smaller males may also be tolerated in the group range. Females may sometimes share day-nests, and there is a suggestion of suckling each others' young. More rarely males nest with each other or with females. Both sexes "urine-wash", urinating onto a cupped hand, then wiping hand on foot and both on branch, which presumably leaves information for others sharing the range.

Coquerels' mouselemur, *Microcebus coquereli* (or *Mirza coquereli*), is a larger animal, weighing around 300 g. It lives in discontinuous patches in dry western woodlands from the Mangoky river northward to the Sambirano region. It is fairly carnivorous — even eating lesser mouselemurs in the accidents of

Fig. 11.1 *Microcebus murinus*, the mouselemur (J.-J. Petter)

captivity. It mainly feeds on the sugary exudates of insects. Coquerels' mouselemur, surprisingly, is monogamous. Male and female join in the second half of the night to shout challenges at neighbouring pairs. They scent-mark with urine, faeces, and saliva. They build spherical nests of twigs, measuring some 50 cm in diameter. They mate and give birth once a year during austral summer. The twin juveniles play for hours with each other as they grow up, behaviour which is taken up again by the mated adults, who nightly hang by their feet, play-wrestle and groom.

The dwarf lemur has two species: *Cheirogaleus major* (340 – 600 g) in the eastern forest, and *Cheirogaleus medius* (220 – 400 g or 1778? in another source) in the south and west. They mainly eat ripe fruit and flowers. They rarely make twig or leaf nests: instead they spend the day in hollow trees. Little is known of their social life, but they mate in October-November at the beginning of the rains, and give birth to two or three young. Gestation in *C. major* is 70 days. They mark trails and ranges by dragging faeces along a branch. The most remarkable fact about dwarf lemurs is that they aestivate

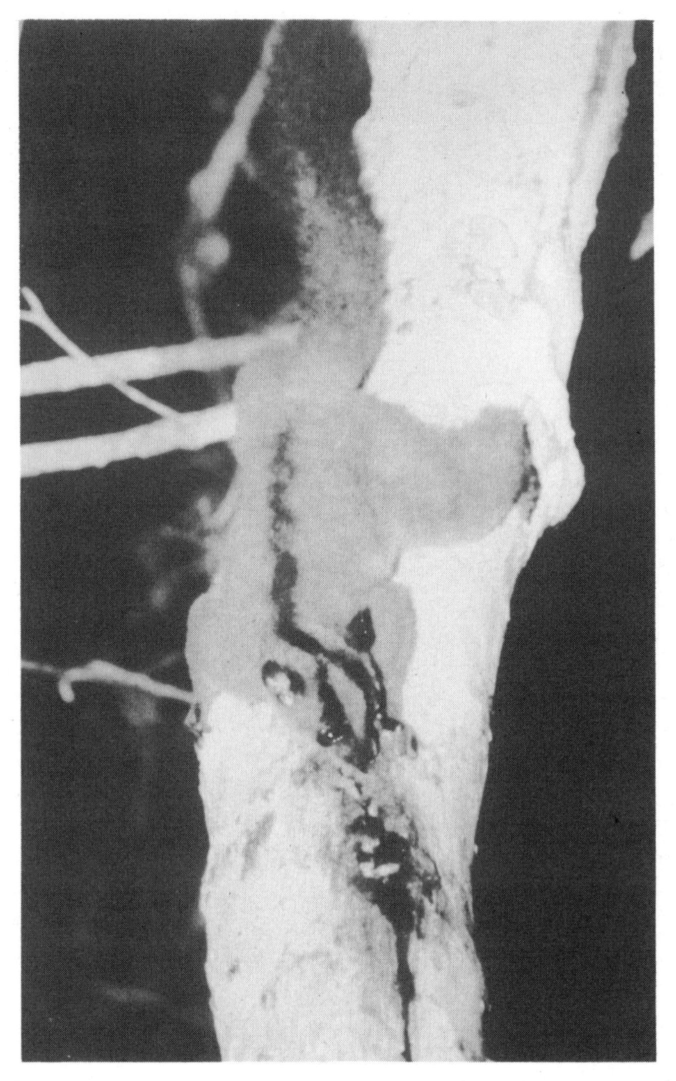

Fig. 11.2 *Phaner furcifer*, the forked lemur, feeding on the gum of *Terminalia mantaliopsis* (C.M. Hladik)

during the dry season living on fat stored at the base of the tail. Malagasy say that they sometimes aestivate in tunnels 50 cm long, following roots deep in the ground. They are often found in galleries inside living rather than dead trees which provide a humid microclimate.

The hairy-eared dwarf lemur, *Allocebus trichotis*, is known from just four specimens collected in 1875, 1880 and 1965. It resembles a mouselemur, but with larger hands, feet and tail. Its teeth are more like *Phaner*'s which may mean that it eats a diet of tree-gums. The 1965 specimen was captured in the eastern forest near Mananara.

The forked lemur, *Phaner furcifer*, is a more golden brown than the other Cheirogaleids, with a deep black-brown stipe down the back which forks on the forehead into a stripe to each eye. Adults weigh 360 – 500 g. They live in western woodlands from the extreme north, in the Montagne d'Ambre, southward to the Onilahy river. There is an isolated population near Amboasary in the southeast, and another in the eastern rainforest of the Masoalala peninsula. Forked lemurs travel high in the woodland canopy, on both large and small branches. They specialize in a diet of exuded gums and insects, which allows them to remain active through the dry season. They are more visually oriented than other Cheirogaleids, emerging earlier at dusk and are sensitive to changes of the moon. Males scent-mark with a throat gland both on branches and on their females. Their most noticeable display, though, are their calls. The male follows a female through the forest with soft contact grunts, which may rise in intensity to loud clicks which the female answers in duet over 50 – 60 m. A still louder squawk is given and answered in "concerts" between rival males. The social structure seems mainly monogamous but at least one male had two females. According to Malagasy woodcutters, *Phaner* mates after the rains begin in November, bears one young, and leaves it in a tree hole only a short time before carrying it on stomach fur then on the back. This would make it rather like a true *Lemur*.

Lemuridae: the true lemurs

The genus *Lemur* was named by Linnaeus for the lemures, luminous-eyed ghosts of the Roman dead. At least black lemurs, the mongoz and the ringtail reached England by the 18th century, and Zoffany painted a ringtail tree-climbing above the elegant Atholl family on the banks of the River Tay.

All *Lemur* are cat-sized quadrupeds, head and body about 50 cm long, and very similar in shape despite their variety of colours. They have long muzzles held downwards so their eyes stare forward above, hindlimbs longer than forelimbs, and plumed or bushy tails longer than head and body.

The ringtail, *Lemur catta*, is the best known. It is pale grey above, white below, with black bandit eye-spots and a black-and-white ringed tail, and weighs about 2.8 kg. It is found throughout southern Madagascar to somewhat north of the Mangoky river. Like other *Lemur* it eats fruit and young leaves. Unlike other *Lemur*, it is semi-terrestrial, promenading through much of its day-range on the ground, tail in a pert question-mark over its back. The males have scent glands near their armpits and a large glandular patch, accompanied by a horny spur, on each wrist. They gouge the scent into saplings with their spurs. They also draw the tail between their wrists, then shiver the tail at opposing males during prolonged "stink fights". They have a wide variety of calls: purrs in close contact, clicks and mews as the troop bounces through the trees, meows if separated from the others. They yap in chorus at ground predators, but scream, streaming downward in the trees, if attacked by a hawk. Males sing from troop to troop with a howl like a coyote's.

They live in troops of about 12 – 25 animals, including several adult males. Males may change troop in November, after a period of wandering and stink fights. They have a strict male hierarchy for most of the year, but females take priority over males for food, space, and in their rare threats. Mating is synchronized, lasting perhaps only two weeks in April. (Timing may vary from forest to forest.) In the laboratory timing is set by decreasing day-length; in the wild a two-month crescendo of calls, spats, and stink-fights may help synchronize mating. Male hierarchy breaks down into true fighting as females reach oestrus, which lasts less than a day for each female. Then the troop calms down until the birth season in early September.

Females bear one young; rarely twins. The infant at first rides longitudinally on the mother's belly fur. They are precocious, climbing onto the mother's back, or onto other females, in the first week of life. This is commonly true in terrestrial monkeys, where babies do not have far to fall. Infants mature to juveniles that romp with each other and eat solid food within three months, but do not reach sexual maturity until 2.5 years in the wild.

The black lemur, *L. macaco*, is somewhat sexist in name. Males of the common race are deep brownish black, but females are golden-brown with white beard and ear-tufts. Black lemurs weigh about 2.4 kg. They live in Nosy Bé and the Sambirano. A second race, *L.m. flavifrons* has even blacker males and paler gold females, without ear-tufts. Some may survive in a localized area in the southern Sambirano.

Fig. 11.3 *Lemur catta* (R. Albignac)

Black lemurs live in small bands of 7 – 9 animals, where the males outnumber the females, a condition virtually unique in primates. It seems, as in ringtails, that females dominate males. They mate in April or May and give birth in September (Petter, 1962).

The brown lemur *L. fulvus* weighs about 2.5 kg. It has seven races which ring the island, and are even found on the island of Mayotte. The commonest, *L.f. fulvus*, is grizzled brown of body with black face, the male's beard somewhat whiter than the female's. This race is oddly distributed — some in the mid-section of the eastern forest, some in the dry northwest. Perhaps they ranged across what is now bare plateau when forest connected the two areas. Another race, *L.f. rufus* is similarly divided between humid eastern forest, and dry western woodland across several hundred miles of savannah. These have grey males with red manes and grey patches above the eye; their females are rufus bodied with white clown eyebrows. The races still further north and south are also sexually dimorphic. In the southern reaches of the humid east live *L.f. collaris* and *L.f. albocollaris*, rich chestnut brown with respectively orange and white beards in the males. Further north in the eastern forest *L.f. albifrons* is grey or even black, with snow white mane and chest in the males, and *L.f. sanfordi*, in the Montagne d'Ambre region is grey with white-headed males whose spiky white beards and ear-tufts recall the black lemurs. Finally, *L.f. mayottensis* looks much like *L.f. fulvus*, and may have reached Mayotte fairly recently as pets in someone's pirogue. All these races replace each other geographically, as they do the black lemurs. Black and brown lemurs are good species however: their chromosomes are incompatible and hybrids are infertile.

One of the brown lemurs has been closely studied in the wild: *L.f. rufus* on the west coast. They lived in small bands of about 5 animals, usually with more than one adult male. Up to 75% of their food was tamarind leaves — perhaps the most specialized diet of any wild primate. Their lethargy, leaf-eating, and tiny, overlapping 6 ha home ranges allowed them to attain a density of 1000 animals per km^2 of tamarind forest. They give birth in September, carry their infants transversally across the stomach, and have much slower development than the ringtails, since the infant rarely transfers to other animals, or to the mother's back before 1 month of age, and does not travel independently until 3.5 – 4 months.

A second study of brown lemurs provides a corrective view. A troop of 12 *L.f. fulvus* on the sandy plateau of Ankarafantsika had a home range of 100 ha. This area is natural deciduous forest with no tamarinds, a tree probably originally introduced by Arab sailors. *Lemur*, like other primates adapts its ranging and social structure to its environment, and may have evolved for the most part in much harsher conditions than the lush tamarind woods.

The mongoose lemur, *L. mongoz*, can be distinguished from brown lemurs because its muzzle is greyish-white not black, and is stubbier in shape. It only weighs about 2 kg. Females have grey heads and pure white cheeks. Some males resemble females, others have orange caps and cheeks. They live in northwest Madagascar and on the islands of Anjoan and Moheli. They seem to live in monogamous pairs on Anjouan but usually in small groups of one or two males and two or more females elsewhere. This is strange among primates — only mongoose lemurs, leaf-monkeys on the Indonesian Mentawai islands, and human

Fig. 11.4 *Lemur macaco*, black lemur female, aggressive expression (R. Albignac)

beings have different local populations which are both monogamous and polygamous. It is hard to see anything else the three species have in common. Mongoose lemurs also seem to switch from crepuscular to nocturnal to diurnal habits in different places and seasons.

The crowned lemur, *L. coronatus*, looks like a mongoose lemur with pale grey fur and orange cap in both sexes and also weighs about 2 kg. Its calls, however, somewhat resemble brown lemurs'. It lives in the extreme north in the dry sunlit forests around the Montagne d'Ambre, as well as in the humid growth on the mountain itself, where it overlaps with the brown race *L.f. sanfordi*.

Redbellied lemurs, *L. rubriventer*, are among the most beautiful, and the rarest and the least known of all lemur species. The fur is rich reddish chocolate and tail black. Males stomachs are pale red brown, females pure white. Their rounded close set ears and a white patch at the inner eye corner give their face a different aspect from the browns. They live at high altitudes, and very low density in the eastern rainforest. Troop size is reported at 3 – 5.

Hapalemur is translated in texts as the gentle lemur, but we have rarely met anyone who actually called it that. It is not common enough to have an English common name. Besides, it bites. It is an equivocal animal, usually classed with lemuridae, but transferred by Tattersall to the Lepilemuridae. Hapalemur live in and feed on thickets of bamboo or else in reed beds, leaping from one vertical stem to another

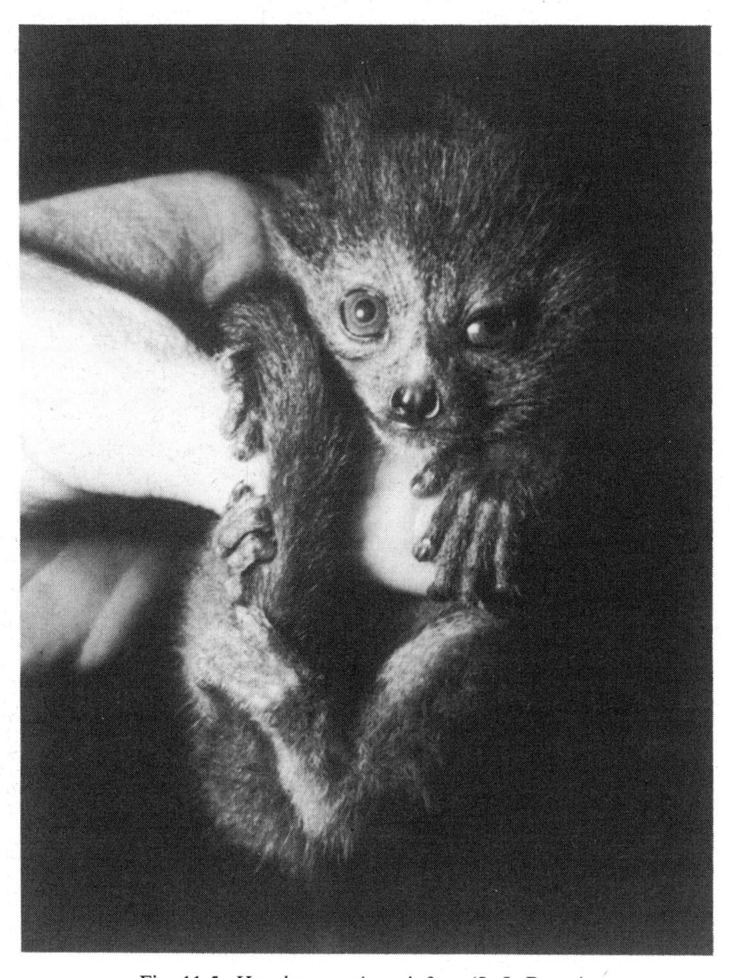

Fig. 11.5 *Hapalemur griseus* infant (J.-J. Petter)

Fig. 11.6 *Hapalemur griseus* mother and young (J.-J. Petter)

with the body held vertical. The commoner species, *H. griseus*, has three races. *H.g. griseus* lives in the eastern forest. Its head and body are about 25 – 30 cm long, weighs about 830 gm, and is a uniform grey colour. It is a crepuscular animal, neither fully diurnal nor nocturnal. Males have specialized scent-marking glands on armpit and wrist, like ringtailed lemurs. Females mark with genitalia and urine, and both sexes deposit faeces in particular places unlike *Lemur* species. Perhaps they have a territorial ownership of particular bamboo clumps established by scent-marking. Their groups are small, usually 3 – 4, and their intolerance of each other in captivity suggests that they are monogamous. The female gives birth to one young in December-January after a gestation of 140 days (12 days longer than the larger brown lemur). She may normally leave it for a time in a hiding place for it is born fully furred but with a weak grasp. In the one observed birth the mother transported the infant in her mouth, but mainly defended its nest box, even from the male. However, groups of 30 have been seen in *H.g. occidentalis*, the western form, discovered in a tiny area near Antsalova by George Randrianasolo in 1972. The third race, *H.g. alaotrensis* was found by J-J Petter in 1968. It is larger than the first two (head and body 40 cm) and darker brown of fur. It leaps among the *Phragmites* reed beds of the great Lake Alaotra, has groups which are small at one season, but congregate into 20 or 30 in another. It escapes predators by jumping in the water and swimming away. The commoner *griseus* may be increasing in numbers as cut forests regrow as bamboo. The Aloatrà form has a bleak future as the lake is steadily drained for rice irrigation, the reed beds are burnt yearly, and hapalemur which flee the fire end up in the stew pot.

The other species, *H. simus*, was first described in 1870, then found after a century's lapse by Petter

Fig. 11.7 *Hapalemur griseus* is a specialized bamboo feeder (J.-J. Petter)

and Peyrieras in 1972. It is still larger, with dark brownish fur, and a distinctive throat gland in the male. It coexists with *H.g. griseus* in a tiny area of eastern forest near Mananjary, and may live in slightly larger groups than *griseus*, of 5 – 6 animals.

The variegated or ruffed lemur, *Varecia variegata*, was traditionally classed as *Lemur*. It seems so different as to merit its own genus, however. Its colour variants are also a muddle. There is one clearly distinct form, the red ruffed lemur, *V.v. rubra*, with black face and limbs and red everywhere else except a white oval at the nape of the neck. This lives only on the Masoalala peninsula. The other race, *V.v. variegata*, lives the length of the eastern forest from Maroansetra south to the Mananara river. It has a gamut of patterns, all piebald black and white with black limbs and faces and white chin and ear-tufts. The two races hybridize in captivity, and perhaps in the wild. The ruffed lemurs you see in zoos are likely to be patchwork black-grey-white-dilute orange, neither the crisp black and white nor the glowing rufus of their forebears.

They are larger than *Lemur* — 60 cm head and body, weight about 4 kg. They have wholly different calls including a group roar that carries at least 1 km through the forest. Males have a throat scent gland. They bear twin young, which are sparsely furred and small. A captive female constructed a nest of some 30 sticks before giving birth, though Malagasy say the young may also be left in a tree-crotch or bracket

Fig. 11.8 *Hapalemur simus* was rediscovered after 100 years (J.-J. Petter)

epiphyte. The young remain mainly in the nest for three weeks, and are never carried on the mother's fur. Group sizes in the wild are small, only 2 – 5 animals. *Varecia* is little known, and may remain so — it is highly edible.

Lepilemuridae: the link

Lepilemur may not be worth a family of its own. On most anatomical measures, however, it comes out intermediate between Cheirogaleids, Lemurids, and Indriids, as distant from them as they are from each other. If it does not stand apart, it is hard to know which group it belongs to. In Tattersall's classification Lepilemur and Hapalemur resemble Eocene Adapids, and thus form an early, distinct lineage. lineage.

Lepilemur is a little grey beast, hard to sort out as Peter Scott's little brown birds. Its head and body are 25 – 35 cm, weight 500 – 900 g. Texts call it the "sportive lemur" but lepilemur seems as good a

Fig. 11.9 *Lepilemur* in its daytime shelter (J.-J. Petter)

common name. Like Cheirogaleids it is nocturnal. Like hapalemur and the Indriids it eats leaves, not insects, and moves by vertical clinging and leaping between vertical stems. It is distributed round Madagascar in a nearly complete ring. There are seven forms which differ so much in chromosome type as well as anatomy they are probably good species: *L. mustelinus* in the northeastern forests south to Tamatave, *L. microdon* in the central and southeastern forests from Tamatave to Fort Dauphin, *L. leucopus* in the spiny desert of the south, *L. ruficaudatus* in the southwest from the Onilahy river to the Tsiribihina, *L. edwardsi* in the north west, *L. dorsalis* in the Sambirano, and *L. septentrionalis* in the Montagne d'Ambre. This last was discovered as a chromosome count on a microscope slide. Only as Rumpler realised it could not hybridize with its neighbours did Jungers confirm its distinctive anatomy.

Lepilemur in the spiny desert have the ancient primate pattern of harem-by-location. The male announces his status by squawking at his neighbours from a high perch on an Alluaudia spire. They bear one young in October which the mother may leave "parked" on branches while she feeds, rather than in her tree-hollow or nest. She transports the young infant in her mouth, though older juveniles sit hugging her back or stomach. The oddest aspect of lepilemurs are their leaf-eating adaptations. Leaves are a high-bulk, low-energy food, so most leaf-eating primates have a large body size to accommodate their large gut. Lepilemur are the smallest primate leaf-eaters and one of only two nocturnal ones (the other is avahi — see below). They may increase their digestive efficiency by eating some of their own

Fig. 11.10 Pair of*Avahi laniger* (J.-J. Petter)

Fig. 11.11 *Propithecus verreauxi coquereli*, northwestern sifaka (J.-J. Petter)

daytime faeces, which probably allows the hind-gut bacteria to help break down cellulose — rabbits do likewise.

This leafy diet affects their distribution. On the Ankarafantsika plateau, lepilemur live in the sandy areas far from water holes, while fruit and leaf-eating species are concentrated in the low ground near ponds or springs. The leaf-eaters thus can maintain high density on dry soils, and through dry seasons.

Indriidae: the soaring leaf-eaters

All Indriids are specialized leaf-eaters. They move by vertical clinging and leaping, 5 m and more from trunk to trunk.

Avahi laniger, the woolly lemur, is grey-brown with rounded face and a white stripe up the thigh which helps break up its spherical silhouette as it rests in the daytime. Head and body measure 30 – 33 cm, weight is only 600 – 700 g (or 1300 g in another report). One race, *A.l. laniger*, lives throughout the eastern forest, the other *A.l. occidentalis*, on the Ankarafantsika plateau. Since they have similar size and diet to lepilemur it seems likely that the two animals are in competition. In fact, in forests where both coexist, the density of both seems much lower than where just one species is found. Avahi differs sharply from lepilemur by a diet of young, not mature, leaves, which means it is commoner near water, while lepilemur ranges in drier areas. It bears its single young in July-August. This slow-growing baby

Fig. 11.12 *Indri indri*. Indri are monogamous, resembling Malaysian siamang in lifestyle and song (J. Pollock)

clings to the mother's fur like a higher primate, and is presumably not weaned until the wet season, along with the lepilemurs born in October. Avahi are monogamous. Four groups in the eastern forest had sharply marked territories with little overlap. They have long distance whistles which provoke choruses of replies, as well as submaxillary scent glands. However, in the western Ankarafantsika plateau their ranges overlap by about 60%. Like *Propithecus*, their territorial behaviour seems to differ in different forests.

Sifaka, the genus *Propithecus* has two species and some eight subspecies. Jean-Jacques Petter takes his stand that *P. diadema* includes the most beautiful lemurs of Madagascar. They are big animals, weight 5 – 9 kg. *P.d. diadema* lives in the eastern forest, from somewhat below Maroansetra south to the Mangoro river, and perhaps even in the upper ridge of escarpment between the Mangoro and the plateau. Its black face is encircled with a white ruff, its head and back deep brown to charcoal grey, while the sacral area, arms, legs and tail are bright golden-orange. Strangely, the other races can be pure white (*P.d. candidus*, which lives from Sambara into the Masoalala peninsula) or pure black (*P.d. edwardsi* and *holomelas*,

Fig. 11.13 Indri (J. Pollock)

the southern forms which continue to the Mananara river, and *P.d. perrieri* a tiny population in the Analamerana, dry forest on calcareous rock just south of the Montagne d'Ambre). Virtually nothing is known of diademed sifaka behaviour and ecology, except that their group size is usually 3 – 6, and they give birth to single young. They are hunted for food more than any other lemurs.

The other species, *P. verreauxi*, by contrast, is among the best studied and most photographed of wild lemurs. These are the white sifaka of the west. They are smaller than the eastern species, only 3 – 5 kg, with 45 cm head and body length. *P.v. verreauxi* is white with a brown cap and white forehead. It lives throughout the south and up to the Tsiribihina river — apparently so successfully in the semi-desert because it can eat mature leaves and never needs to drink free-standing water. *P.v. deckeni* has pure white fur. It lives in the northwest from the Tsiribihina to the Mahavaly river. *P.v. coronatus* which lives between the Mahavaly and the Betsiboka rivers has an all-black head and neck cape. *P.v. coquereli*, north of the Betsiboka, has an all-white head in contrast, but a maroon eagle-wing pattern on its outer thighs and forearms. All the races are fairly variable in colour. The eagle-wing pattern reappears in all the other races as a melanistic trait, which was thought a separate subspecies. Melanistic animals have dark heads and are chocolate-patched instead of sharing *coquereli*'s purplish tinge.

Sifaka eat mainly young and mature leaves, as well as some flowers, bark and fruit. A sifaka is such a confirmed leaper that when it (rarely) comes to the ground it bounces along on its hind legs like a child in a sack race, arms flailing wildly over its head. They live in groups of 3 – 11, averaging about 5, with several males and several females. Females dominate males, as in other social lemurs. Males have a special scent gland on the throat. They give a roar of alarm to flying hawks, and insult ground predators by approaching and saying shi-fakh! shi-fakh! Other than scent, there is no long-distance signal answered by other troops. In the patchy forest of Ankarafantsika, *coquereli* troops overlap home ranges entirely. In two forests in the south, *verreauxi* defend discrete territories. The territories in spiny desert are twice as large as those in a rich gallery forest. In the spiny bush they apparently rely mainly on scent posts. In the gallery forest troops fight ritualized "battles", a kind of arboreal chess-game in which white knights leap to occupy and mark empty spaces, facing out from their own territories, so opposing champions sometimes land on the same branch back to back. They mate at the end of January with male-male fighting and inter-troop migration. One young is born in late July or early August. The infant clings transversally to the mother's stomach for a month, then rides on her back. It is groomed by all group members, including the males, so far as the mother lets them. Sexual maturity is not reached until 2.5 years or perhaps more.

Indri indri is grandest of all the modern lemurs. Its head and body measure 70 cm, though its stumpy tail is just 3 cm long. Adults look as though they weigh 15 – 20 kg, though the one recorded weight is 6 kg. Indri is black and white — black head, tufted teddy bear ears, black back, hands, feet and the leading edge of thighs, and forearms. Its limbs are mostly white, and an inverted V of white covers lower back and spine. If they did not leap and call so magnificently they could be piebald clowns. An animal whose wails are answered from 3 km away by other indri, or which launches itself lightly over 6 – 7 m gaps, is no clown. Indri live in the eastern forest from north Maroansetra south to the Mangoro river. Although the range is vast, they have very low population density anywhere, and so they are particularly vulnerable to forest clearance. They have never been successfully kept in captivity.

Indri are monogamous. Male and female sing together, and their juveniles join in after a phrase or so. Each indri can probably identify the individuals in other groups by their wails. They are also strictly territorial, challenging others by song and perhaps by salivary scent marks. In their monogamy, territoriality, and song they resemble the gibbons of south east Asia. However, as usual in lemurs, the indri female is dominant. She feeds in the choicest part of the tree while her male feeds below, and occasionally cuffs him if he climbs too high. They bear only one young every three years, in July, and carry it on the fur.

Daubentoniidae: the aberrant aye-aye

No one doubts that the aye-aye *Daubentonia madagascariensis* is worth a family of its own. It took scientists a hundred years even to decide that the aye-aye is a lemur.

It is about the size of a *Lemur*, weight for 1 specimen 2.8 kg, blackish brown, with paler face, and long white-tipped guard hairs on body and tail. It has ever-growing incisors like a rodent and huge mobile ears. Its third-finger is skeletal, not actually longer than the fourth, but seeming so because it is held crooked away from the others as the animal walks. The third finger is too delicate for walking; it is a specialized probe. All the digits except the great toes bear claws, unique among lemurs.

The aye-aye uses its claws to climb large trunks, its bat-ears to listen for insect crepitations beneath the bark, its beaver-teeth for stripping, and its skeleton finger to extract the juicy grubs. Madagascar has no true woodpeckers: the aye-aye fills their niche. Strangely, another such oddity lives in New Guinea: *Dactylopsia*, a marsupial which has all the same adaptations.

The aye-aye is nocturnal and solitary. It builds twig nests, and bears one young in October-November. It has apparently no long-distance calls. The "ha-hay" call is a blowing through open nostrils with closed mouth, probably given to warn or frighten. It marks branches with urine, and has a glandular scrotum.

It is, or was, an inhabitant of the eastern coastal forest, and probably also the forests of the eastern escarpment. Most tribes of the east fear it, believing it forebodes death in the village. This dread, as well as the fact that it efficiently raids coconuts with its teeth and probing finger, means there are few remaining aye-ayes. In 1966, Petter and Peyrieras released nine aye-ayes on the Island of Nosy Mangabe, hoping that the forestry department could protect it as a World Wildlife Fund Reserve. Aye-aye nests

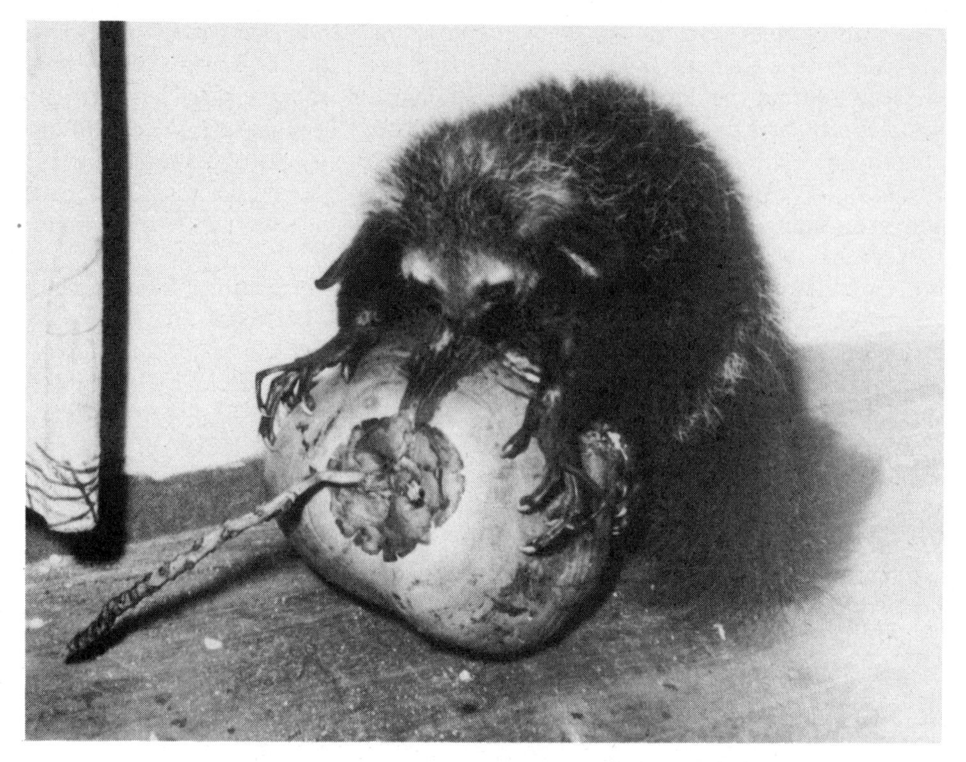

Fig. 11.14 *Daubentonia madagascariensis*, the aye-aye on a coconut. It is hated for its depredations on coconut plantations as well as feared for its evil luck (J.-J. Petter)

and one animal have been found there as recently as 1975, and the island still seems intact from poachers. Elsewhere, however, there is small hope of saving this one of the world's rarest, and strangest mammals. The aye-aye's skeleton finger points, not to villagers' deaths, but to its own.

The extinct giants

All the subfossil lemurs were larger than those that remain. There was *Varecia insignis*, and *V. jullyi*, with skulls 15% larger than the modern ruffed lemur. There was *Daubentonia robusta*, a larger aye-aye. *Mesopropithecus pithecoides* and *M. globiceps* were large indriid cousins of the sifaka, thé size of guenon monkeys.

Others were still bigger, and unlike any of the survivors. *Paleopropithecus ingens* and *Archaeoindris fontoynonti* were vertical-bodied animals with flattened faces and round skulls. Paleopropthecus hung by their hands and clambered more like present-day orang-utans, or perhaps even sloths, than the leaping indriids. *P. ingens* was as large as a female chimpanzee, *Archaeoindris* to judge by the single known skull was bigger still. Two other genera were terrestrial. *Archaeolemur majori* and *A. edwardsi* resembled small baboons in their equal-length terrestrial limbs and their grit-worn teeth. *Hadropithecus stenognathus* was fleeter of foot, long-legged like a patas monkey, and chewed hard seeds like the ancestors of men.

Megaladapis was the biggest of all. *M. edwardsi* may have weighed two hundred kilos; like a big male orang-utan. Other species, *M. madagascariensis* and *M. grandidieri*, were similarly built, if smaller. They were more closely related to lemur or hapalemur than to the indriids.They all had heavy cow-like jaws

Fig. 11.15 The aye-aye in the wild (J.-J. Petter)

and molars, with a gross lemuriform skeleton and hands and feet that could wrap right round a tree-trunk. They probably clung to the trunk like immense koalas, extending head and neck and perhaps a prehensile snout to drag in a leafy branch. If they had to shift tree, they may have bumbled across the ground in frog-like hops — until the first men came.

What happened? There have been long arguments about their disappearance. The climate has not changed significantly, at least not in the last 1000 years since all these species left their bones in what is now bare plateau, hundreds of miles from any forest. It seems more likely that fire, habitat destruction and hunting have already killed off a third of the lemurs. Introduced zebu extended the grasslands, while out competing the grassland fauna. Madagascar was a lost world, preserving and evolving its own community of species, including creatures which now live only as legends. They died because people could not imagine they would die, and did not care.

That was the first wave of extinction in Madagascar. The second is taking place today.

Fig. 11.16 What we have lost. Back row, L to R: *Megaladapis, Archaeoindris, Paleopropithecus, Archaeolemur.* Front row: *Hadropithecus* with mouselemur and indri, smallest and largest of the survivors. All of these genera coexisted at Ampasambazimba, a single fossil site (A. Jolly)

REFERENCES

Charles-Dominique, P., Cooper H.M., Hladik A., Hladik C.M., Pages E., Pariente G.F., Petter-Rousseaux A., Petter J-J, Schilling, A. 1908? *Nocturnal Malagasy Primates.* Academic Press, New York. 215pp.

Dewar, R. in press.

Doyle, G.A. and Martin R.D., Eds. 1979 *The Study of Prosimian Behavior.* Academic Press, New York, 696pp.

Jolly, A. 1966 *Lemur Behavior.* University of Chicago Press, Chicago. 187 pp.

Jolly, A. 1980 *A World Like Our Own*: Man and Nature in Madagascar. Yale University Press, New Haven. 271pp.

Martin, R.D., G.A. Doyle and A.C. Walker 1974 *Prosimian Biology.* Duckworth, London, 983pp.

Petter, J-J. 1962 *Recherches sur l'Ecologie et l'Ethologie des Lemuriens Malgaches.* Mem. du Museum National d'Histoire Naturelle, A: 27 1 – 46.

Petter, J-J, Albignac, R. and Runbler Y. 1977 *Mammiferes Lemuriens (Primates Prosimiens).* Faune de Madagascar 44: 513pp. O.R.S.T.O.M., Paris.

Petter-Rousseaux, A. 1962 Recherches sur la Biologie de la Reproduction des Primates Inferieurs. These de la Fac. des Sciences, Paris. Mammalia, Paris, 87pp.

Richard, A.F. 1978 *Behavioural Variation.* Case Study of a Malagasy Lemur. Bucknell University Press, Lewisburg. 213pp.

Tattersall, I. 1982 *The Primates of Madagascar.* Columbia U. Press, New York, 382pp.

Tattersall, I. and Sussman R.W. 1975 *Lemur Biology.* Plenum Press, New York, 365pp.

Appendix: MALAGASY LEMURS

Latin	English	Malagasy	Region
Family CHEIROGALEIDAE			
Microcebus murinus	gray mouselemur	tsidy, koitsiky, pondiky	W, S
M. rufus	rufus mouselemur	tsidy, tsitsihy	E
M. coquereli	Coquerel's mouselemur	tsiba, tilitilivaha, setohy, fitily	W, Sa
Cheirogaleus major	greater dwarf lemur	tsidy, tsitsihy, hataka	E
C. medius	fat-tailed dwarf lemur	matavirambo, kely be-ohy, tsidy, tsitsihy	W, S
Allocebus trichotis	hairy-eared dwarf lemur		E
Phaner furcifer	forked lemur	tanta, tantaraolana, vakiandrina, vakivoho	W, S, E, Sa
Family LEMURIDAE			
Lemur catta	ringtailed lemur	maki, hira	W, S
L. macaco	black lemur	akomba, ankomba, komba	Sa
L. fulvus	brown lemur	gidro, boromitoko, varika, varikosy	E, W, Sa, C
L. mongoz	mongoose lemur	dredrika, hira, komba	W, Sa, C
L. coronatus	crowned lemur	ankomba, varika	W
L. rubriventer	redbellied lemur	tongona, soamiera, barimaso	E
Varecia variegata	ruffed lemur	varikandana, varikandra	E
V. insignis	variegated lemur	varimena	F
V. jullyi			F
Hapalemur griseus	hapalemur	bokombolo, kotika, bandro, bekola	E, W
simus	broad-nosed hapalemur	varibolo	E
gallieni			F
Megaladapis madagascariensis			F
M. edwardsi			F
M. grandidieri			F
Family LEPILEMURIDAE			
Lepilemur mustelinus	lepilemur	trangalavaka, kotrika, fitilikily, hataka, varikosy	E
L. microdon			E
L. leucopus			S
L. ruficaudatus		boenga	W
L. edwardsi		repahaka, boenga	W
L. dorsalis		apongy	Sa
L. septentrionalis		mahiabeala, songiky	W
Family INDRIIDAE			
Indri indri	indri	babakoto, amboanala, endrina	E
Avahi laniger	avahi, woolly lemur	fotsife, ampongy, avahy, tsara fangitra	E, W
Propithecus diadema	diademed sifaka	simpona	E
P. verreauxi	vereaux sifaka	sifaka	W, S
Mesopropithecus pithecoides			F
Archaeolemur majori			F
A. edwardsi			F
Hadropithecus stenognathus			F
Paleopropithecus ingens			F
Archaeoindris fontoynonti			F
Family DAUBENTONIIDAE			
Daubentonia madagascariensis	aye-eye	ha-hay	E
D. robustus			F

*From Tattersall 1982
E = humid east; W = dry west; S = semi-arid south; Sa = Sambirano; C = Comores Island; F = Subfossil.

CHAPTER 12

The Malagasy and The Chameleon: A Traditional View of Nature

GUY A. RAMANANTSOA

The foreign scientist or expert is often inclined to believe that the Malagasy are indifferent or even hostile towards their natural surroundings. The large scale of forest clearance, bush fire and poaching inevitably shocks such a scientist, who marvels at the unique forms of vegetable and animal life with which he is just making contact.

The rural Malagasy peasant, especially in forested regions, lives in contact and at ease with nature. It is the source of material benefits, and also of inspiration — poetic, moral or even religious. Observation and investigation have led him to precise and far-ranging technological knowledge of nature, as Guillaumet describes in this volume for the Betsimisaraka's use of Eastern Forest plants.

In this article we wish to furnish an example of the close relation between man and wild animals in Madagascar. We have chosen a group which is negligible *a priori*. The Chamaeleontidae are neither edible nor truly dangerous. They might well have passed unnoticed. However, the richness of proverbs relating to them shows, instead, that these animals have provoked the *jeux d'esprit* which result in such proverbs. These formal phrases are lapidary formulas that illuminate orations, and even the everyday conversation of anyone who wishes to be known for his verbal wit and wisdom.

The Malagasy Chamaeleontidae are a popular group in the scientific world. They illustrate to the full those characteristics of "refuge for primitive forms" and "laboratory of evolution" which scientists repeatedly attribute to the Great Island. Chameleons are a heritage of the Cretaceous Era. The family contains only 2 genera, but with 59 Malagasy species, all of them endemic. Genus *Chamaeleo* Laurenti 1768 has 37, while 22 fall in the genus *Brookesia* Gray 1864. These numbers mean that Madagascar is home to 30% of the world's known *Chamaeleo* and 66% of the *Brookesia*. Some scientists even call it "the chameleons' refuge".

The first European mention of the Malagasy Chamaeleontidae dates from 1595. The first taxon, *C. bifidus* A. Brongaart was described in 1800, and the most recent, *B. bonsi* and *B. legendrei* G.A. Ramanantsoa, in 1979. We are convinced that the list of forms will grow still longer. Study of specimens now at the E'tablissement d'Enseignement Superieur des Sciences Agronomiques, of the University of Madagascar, will lead to the description of at least 3 new *Chamaeleo* and 7 new *Brookesia*. Besides, many sites in the field await the first visit of anyone collecting chameleons.

The largest *Chamaeleo*, *C. oustaleti* F. Mocquard, 1894, and the smallest, *C. nasutus* A. Dumeril & Bibron, 1836, live in Madagascar. Similarly, the largest *Brookesia*, *B. perarmata* (F. Angel 1933) and the smallest, *B. minima* O. Boettiger 1893, are also Malagasy forms. The maximum recorded lengths for these species are:

C. oustaleti	68.5 cm
C. nasutus	0.8 cm
B. perarmata	11.0 cm
B. minima	3.2 cm

It seemed interesting to ask how these Chamaeleontidae, so captivating for the scientist, are seen by Malagasy. We therefore decided to study chameleons within the framework of a general ethnozoological inquiry on the vertebrates of Madagascar. We attempted, by interviewing villagers, to collect vernacular names, proverbs, sayings, stories, legends, popular beliefs and songs which refer to the Chamaeleontidae.

The richness of the harvest greatly exceeded our expectations:

65 names, of which 37 previously unrecorded,
76 proverbs and sayings of which 67 unrecorded,
4 stories and legends of which 3 unrecorded,
8 "jijy", or improvised poems, all unrecorded,
4 popular beliefs, 3 unrecorded.

This list is far from complete, for we could not make inquiries in depth in each of the 18 tribes.

This rich repertoire reveals the special interest which Malagasy share in the Chamaeleontidae. But what is the nature of this interest?

First of all Malagasy feel repulsion when faced with chameleons:

— Ratsy karaha Kandrondro (Tsimihety tribe)
Ugly as a chameleon
— Tsatsaka no i Taha
Ka mifaninjy fa samy raty (Mahafaly tribe)
Between *Phelsuma* and chameleon
Don't quarrel, you are equally ugly
— Kanondro mitety letra
Mitera fatsy tsara (Tsimihety)
Chameleon promenading on a rock:
Proud but no beauty.

People deduce the chameleon's "pride" from its mannered locomotion. If it is not frightened it rocks back and forth several times before putting down each foot.

This Malagasy reaction contrasts with the usual European one, which is to find chameleons beautiful with their amazing changeable colours.

For people with more delicate sensibilities, the chameleon is not merely repulsive, but horrible. Such people claim to "sense" them at several metres distance, and make detours to avoid such an evil encounter. Other villagers warned us never to tease such people by showing them one of our "proteges" — they would be capable of uncontrolled violence. . . . .

Retrama, a healer of the village of Ampototse in the South, takes advantage of this fear in treating mental disorders with a kind of shock therapy. First a patient remains with him drinking infusions of calming medicinal plants. At the end of his stay, he is led to the river, where the traditional practitioner forcibly immerses him three times in succession. While he is still suffocated, Retrama thrusts at his face a large chameleon, *C. verrucosus* G. Cuvier 1831 which is abundant in the region. The client who reacts sensibly by taking flight, is considered cured!

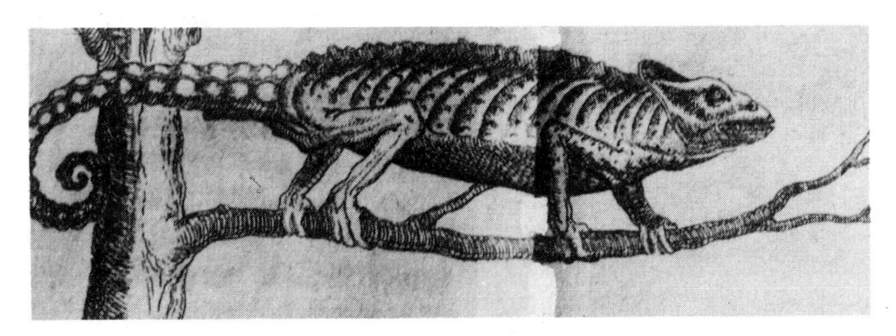

Fig. 12.1 The chameleon (Megiser, 1609)

This horror or revulsion results, in our opinion, directly from the chameleons' own morphology. The fear comes from its means of capturing prey. Decary reported in 1950 that on the east coast women fear the animal because "it could put out an eye, so they would never be able to find a husband". On the High Plateaux, chameleons are said to "spit" in one's eye, which brings on a frightful malady that finishes by depigmentation of the iris. . . . It is true that chameleons project their tongues to a great distance, to catch insects. It is now proved scientifically that they are stimulated by the prey's movements. It is not impossible that eye movements could also attract them, in which case the gluey tip of the tongue would result in a very disagreeable situation for the human being.

Besides this physical fear, there is also a moral fright. Decary reports that among some tribes the chameleon portends ill-luck. Among the Mandiavato whoever encounters a chameleon must expect misfortune in his family. In the Bezanozano it is only if you crush one that you need be uneasy. In the south, we learned, it is the corpse that must be avoided. A living animal is a good sign. It is even in one's interest to stride over it. In the Maroansetra region, we learned as well that the year will have violent cyclones if the chameleons climb to the tops of the trees, for they normally stay in the lower levels.

The *Brookesia* cause mortal fear, everywhere and always. Even their names are significant: Andro[1], Tsiny[2], Nala[3], Ramilaheloka[4]. "Andro" is a synonym for ill-luck. "Tsiny" and "Nala" are genies of the forests, supernatural powers which are consulted about various problems of life by way of tranced mediums. "Ramilaheloka" literally means "He who seeks to condemn, or he who seeks to make guilty". In fact, popular belief affirms that whoever crushes a *Brookesia* will suffer ill-health which might even end in death.

Here are a few proverbs expressing the fear of the Brookesia:

"Mahatsidia vokon' Anjava[5]
Kely izy fa mafoaka
Step on a knot of *Brookesia*:
A small animal but expect a great misfortune

Aza mairano Renimilaheloka[6]
Don't behave like the *Brookesia*

This second proverb is addressed to anyone who only thinks of doing evil without acting. He may cause harm like the Brookesia which makes anyone ill that steps on it, even if the act was involuntary.

[1]Bara Harongana; [2] Sakalava du Boina; [3] Tsimihety; [4] Merina.
[5]Tismihety; [6]Northern Betsimisaraka;

Aleo mahatsidia Zanahary[1]
Toy izay mahatsidia Ranovary
It would be better to trample a divinity
Than trample a *Brookesia*

Malagasy ethics forbid harming any animal without valid reason. All animals have their spirits, capable of avenging them. Certain sick people are generally called "Nangarim-biby", the "animal-spirit victims". Note that the kind of animal is not specified in this phrase. Even though one can say that Malagasy respect all animals in this general sense, few of them frighten him so much as the *Brookesia*.

In our opinion the somewhat exaggerated fear of Brookesia results from the risk of crushing it inadvertently, and thus from unconsciously harming it. This has created a kind of collective psychosis.

One might also think that the animal, so tiny and hidden in the humus, is a kind of "trap" which nature sets for man; that nature harbours supernatural beings which are good or bad in relation to humanity. This fear results from associating *Brookesia* with wood-genies. Its names are witness: "Zanak'-anjava", "Zanaka-tsiny" (little ones of the genies). From there, we arrive at the confusion we have already seen with the names "Tsiny", "Nala" (genie). This confusion is inevitable once one has attributed the same evil deeds to Brookesia as the bad genies themselves.

In Sakalava country, on the island of Nosy Bé, Brookesia is feared as everywhere else, but is also the object of particular respect, and a debt of thankfulness. A legend tells that it once saved the life of an ancestor from the attack of a cannibal monster.

Like all people who basically live from the products of the earth, Malagasy are attentive observers of nature. They are struck by the ethological "monstrosities" of the chameleon: eyes that swivel through 180° independently of each other, prehensile feet and tail (though *Brookesia*'s tail is not wholly prehensile), colour changes (these are largely changes of mood, not camouflage), stalking prey at a distance. . . . The chameleon's extraordinary behaviour has inspired most of the proverbs, stories, legends and poems. For example, 86% of recorded proverbs are clearly derived from it.

One describing the visual system is among the most popular. It is known in all tribes and by adults but also children, to whom people like to repeat it. A President of the Republic even quoted it in his speech of investiture:

Ataovy dian-tana
Jerena ny aloha
Todihana ny afara[2]
Behave like the chameleon:
Look forward and
Observe behind.

Without turning its head, the animal indeed directs one eye forward (on the future), and one eye behind (on the past). The proverb counsels prudence. . . .

A lesser-known proverb again makes the chameleon symbol of the same virtue, but this time referring to the limbs which secure it to its support, sure not to fall:

[1] Northern Betsimusaraka.
[2] Merina

Tanalahi miani-kazo
Ny tanana mamikotra
Ny rambo misafelika[1]
Chameleon climbing on a tree:
The feet cling
The tail holds fast

Malagasy interest in chameleons is limited to curiosity, linked to the sentiments of repulsion and fear. This means that the Chamaeleontidae's future is not threatened directly by man in Madagascar. The real menace is from habitat destruction, as for most of the other Malagasy animals. But we should hardly expect massive chameleon collecting for traditional medicine, as is true in Morocco (J. Bons, 1960), or even collecting for peculiar tourist souvenirs, which was the case in Reunion where the Mayor of Saint-Paul had to issue a municipal edict against gathering the "endormi", *C. pardalis* (R. Bourgat, 1969).

There are very few uses of these animals which involve killing them, in Madagascar. Ashes of female chameleons are employed to treat infantile convulsions. Blood collected on the feet of a chameleon is used to make an amulet, supposed to increase a wrestler's agility when he tries to seize his adversary, like the chameleon which clasps its support with such precision.

Finally, chameleons are used to make another amulet which is worn by accused persons on the day of judgement. The vernacular name "Tana" is the root of the verb "Mitana", which means "hold back". One may suppose that the animal "holds back" a judge tempted to hand down a harsh verdict!

The few amulets and cures are made from *Chamaeleo* species. No one ever touches a *Brookesia*.

BIBLIOGRAPHY

Bons, J. et N. 1960 Note sur la reproduction et le developement de Chamaeleo chamaeleon (L). *Bull. Soc. Sc. mat. phys. Maroc.* 40, 4: 323–35.

Bourgat, R.M., 1969 Recherches ecologiques et biologiques sur le *Chamaeleo pardalis* Cuvier, 1829 de l'Ile de La Reunion et Madagascar. These doct. Sci., Montpellier.

Brygoo, E.R., 1971 Reptiles Sauriens Chamaeleonidae. Genre Chamaeleo, *in* Faune de madagascar, vol. 33, Q.R.S.T.O.M. et C.N.R.S. edit, Paris.

Domenichini-Ramiaramanana, B. 1972 Ohabolanany ntaolo, exemples et proverbes des anciens. *Mem. Acad. malgache* 54, 654p.

Ramanantsoa, G.A. 1974 Connaissance des Cameleonides communs de la Province de Diego-Suarez par la population paysanne. *Bull. Acad. Malgache*, 51, 1: 147–49.

Ramanantsoa, G.A. 1979 Description de deux nouvelles especes le *Brookesis*:*B. legendrei* et *B. bonsi*. *Bull. Mus. natn. Hist. nat., Paris, 4ᵉ* ser., No.3, Zool.: 6850–693.

Ramanantsoa, G.A. (en preparation). Observations preliminaires sur lethno-zoologie et l'eco-ethologie des Cameleonides malgaches.

[1]Betsileo

CHAPTER 13

Malagasy Economics and Conservation: A Tragedy Without Villains

ALISON JOLLY and RICHARD JOLLY

In the early 1980s sub-Saharan Africa is reeling under the impact of inflation and world recession, often compounded by debt and drought. Economies which were weak before are now in a state of declining production and more steeply declining per capita income. The 1982 World Bank Annual Report stated that the plight of rural Africa is the world's most serious development problem. Madagascar is no exception. The economic pressures on Madagascar's rural population are more serious than at any time since Independence. The resultant squeeze not only keeps the present generation under pressure, but is forcing them to sacrifice the future. The struggle to survive leads them to ever-increasing use of their one free resource: the forest lands.

Madagascar today has nine million people, with a population growth of 2.6% annually (1975 – 80 figures). Well over three quarters of these people live by agriculture, which contributes about half the GNP (gross national product). Per capita income in 1979 was about $290 with an average decline of 0.4% per year since 1960. The country apparently has a low population density, only 15 persons/km^2. Most of the land, however, is nearly sterile — once forested, now the far end of a fire climax, growing Aristidia bunch-grass which is useless to man or beast. There was only about an acre of arable land per person of the agricultural population in 1975, which is small for an African country, though average for the paddy-rice nations of Southeast Asia. Other estimates, however, suggest that a mere 15% of the arable land is cultivated. It would certainly be possible to increase land use and yields, but the pockets of good land are so dispersed and fragmented that much of it is uneconomic for market crops.

The history of land use in Madagascar has long been one of forest clearance — as in America, Europe, and Africa. The Malagasy immigrants cleared most of their country in the wave of settlement and expansion which led to extinction of the megafauna between 500 and 1700 A.D.

Three decades ago, Guichon estimated that 21% of Madagascar land area was still under forest cover, based on an aerial survey chiefly made in the late 1940s. For comparison, this is about the same percentage as in France today, which the visitor sees as fields, farms, and cathedrals. The eastern seaboard states from Maine to Virginia have almost as much area as Madagascar, seven and a half times Madagascar's population — and twice as much forest. More significantly, the forest lands of Europe and America are roughly stable, since we now appreciate their value and can afford a long-term view. In Madagascar, the forest continues to disappear rapidly. We have no Landsat-based data for more recent estimates than the 1945 – 50 survey. It seems, however, that the forest which would yield to random burning or under the unfamiliar pressure of cattle grazing has long since gone. The remaining forests are already relicts, preserved by the accidents of terrain hostile to cultivation, grazing and fire. The modern forests must be actively felled or grossly overstocked to disappear. This is now happening at ever accelerating rates.

The modern pattern of peasant agriculture differs, like all other forms of life, by biogeographic regions. In the central highlands, for centuries cleared of forest, people chiefly grow paddy rice. There are some river valleys of both west and east, particularly the Lake Alaotra basin, where a constant water supply and fertile soil also allow irrigated rice cultivation. In the eastern and Sambirano rainforests there is shifting cultivation of mountain or dry-land rice, manioc and other root crops. This means felling the forest, then abandoning it to regrow as "savoka", a second-growth thicket which rarely, if ever, has time to return to mature rainforest.

Cattle are the other source of pressure. There are at least 10 million zebu, 2 million goats and sheep, and 1 million pigs in Madagascar, chiefly in the west and south. The zebu are valued for prestige in life and splendid sacrifices at funerals. It is their numbers, not their quality which matter to their owners, reflecting both traditional attitudes and the feebleness of the market economy. The remaining deciduous woodlands of the west often burn at the edges when grass-fires catch in the trees. It has been estimated that a third of Madagascar burns each year, as people set fires for a "green bite" for the herds in the dry season, and as political protest. Some shifting cultivation also remains, though less than in the humid east.

The southern spiny desert is paradoxically too dry to burn. Fires cannot spread in the sparse leaf litter, or attack the succulent plants. However, herds of zebu and goats graze on bark wherever they can. Where there are roads, the spiny desert is also being felled for charcoal and house timbers. There is little cultivation, chiefly maize in the few regions of useable soil.

Madagascar has seen commercial logging in the past, especially for rosewood, ebony, and other marquetry timber. However, the quantity of hardwood per hectare is relatively low, especially compared to the Indonesian rainforest, so the forest has been little exploited for profit. The major threat instead is smallscale farmers exploiting it for their own survival.

Fig. 13.1 Roadside charcoal burners' village in the spiny desert. The forest is cut for cooking fuel for the towns (A. Jolly)

Fig. 13.2 Zebu and goats prevent regeneration. Overgrazing can reduce the semi-arid flora to bare earth (A. Jolly)

Fig. 13.3 The western grasslands are burnt to provide a green bite for the herds in the dry season (A. Jolly)

Fig. 13.4 The eastern forest is felled for shifting agriculture (A. Jolly)

Madagascar's economy is surprisingly fragmented in the modern sector. Unlike many developing countries, it does not depend on a single crop or export. Madagascar exports coffee, vanilla (90% of the world market), cloves, sisal, tobacco, pepper, cotton, sugar — produce of both humid and semi-arid zones. Each region of the island has its own economy. For economics as well as natural history, Madagascar seems more like an archipelago than a country.

Madagascar's diverse regions are linked by an abysmal road system. Inadequate even in earlier times, it was devastated by five cyclones during the summer months of 1981 – 82 which caused an estimated $250 M worth of damage. It is only intermittently possible to travel by 4-wheel drive vehicles between the capital, Antanarivo and the chief port, Toamasina (Tamatave). One train a day links capital to port, on an aging railroad bed. The 515 km "main" road to the southern town of Toliary (Tulear) can be negotiated in the dry season without 4-wheel drive, but at an average speed of 20 km/h. In the wet season it is frequently impassable. This timing does not allow for the almost inevitable breakdowns. There are few or no spare parts; cars and buses are repaired by pirating parts from still more decrepit vehicles, which in turn reduces the available stock. Madagascar at present has no traffic jams.

These "main" roads are being partially reconstructed, one with West German aid, one with Chinese. They give an index, however of the state of the subsidiary roads. Many are simply unuseable, with broken

Fig. 13.5 The forest, once felled, does not return. Clearings may sprout second-growth thicket, or erode, as here, to grassland and bare rock. In the foreground, a fire-killed tree with termite nest (A. Jolly)

bridges and roadbeds eroded into ravines. What this means is that peasants who used to sell cash crops to a passing trader now walk to town with their coffee in baskets on their heads — or else withdraw from the market economy.

Simultaneously the price of those manufactured goods which peasants might buy has skyrocketed, while distribution problems mean such goods are often unavailable. Inflation was officially estimated at 31% in 1981; allowing for the black market it is much higher. There is little soap, most of it too expensive to purchase. Peasants who bought soap five years ago now pound aloe roots with their clothes. Rural clinics lack drugs, rural schools lack writing paper. The national match factory has closed; matches are once again imported from Sweden and China. Flashlight and transistor radio batteries are scarce. Even the local brewery has curtailed production because it cannot get enough bottle caps. There are no substitutes for breast milk, which probably saves many children's lives by forestalling diarrhoea, but kills a few whose mothers cannot feed them.

Madagascar had no growth in real gross domestic product between 1970 – 78, and an estimated fall of 9% in 1981. (For comparison, the U.S. GDP fell 1% in 1981 which felt like recession even there.) The balance of payments has grown worse. The average price of Madagascar's exports rose 12% between

1978 – 81, while the average price of imports rose 50%. As President Ratsiraka pointed out in an October 1982 press conference, one third of the coffee crop in 1972 paid for Madagascar's oil imports. In 1982, 100% of the coffee crop will not pay for oil imports. They would need to run three times harder to stay in the same place.

These large-scale effects rebound on the people who make up the poorest levels of the market economy. The government's increasing burden of debt led the IMF to reschedule debts in both 1981 and 1982, and to extend special drawing rights. It imposed conditions, however, especially the withdrawing of food subsidies. The official price of rice, the staple food of the towns, rose 18% in May 1981 and 100% in May 1982. This applies to the 250,000 tonnes of imported rice, which is roughly the requirement of the urban population. Locally-grown rice, of much better quality, sells for three times the official price in what are politely called "parallel markets". Thus, the poor urban labourer, a high proportion of whose family budget goes to buy rice, is hardest hit by government debt.

Subsistence farmers are probably faring better by comparison. Those peasants whose villages lie on footpaths, not roads, and who have always bought and sold little are not much worse off than before. Madagascar's human statistics reflect this, although it is difficult to be sure of figures for remote areas. The infant mortality rate of 71 per 1000 is the best of African low-income countries, though it might be somewhat under-reported. Death rate of children 1 – 4 is 25 per 1000, life expectancy at birth 47 years (both figures for 1979). These are close to the low-income country averages. So much of the rural economy is highly traditional that there is relatively little malnutrition. The estimated 2486 calories per person per day in 1977 was the highest of African low-income countries.

What the economic crisis does mean for the rural population is that there is no way out to the towns. In fact, reverse migration to the countryside might be occurring with the present lack of employment.

This in turn puts more pressure on the land. Although it would be possible to increase yields on presently cultivated land, this would need investment and fertilizer. With traditional methods of cultivation, Malagasy know the valley farms are occupied, infertile, or both, so they need new clearings on the forest

Fig. 13.6 After the fires, only this remains (Ph. Oberlé)

slopes. They value their zebu herds for numbers of scrawny cattle. Unless they change their ways of cultivation, which means more access to credit and to markets, there is real pressure on land, and forest — all the natural habitat so lovingly described in this book.

In the long term, Malagasy agriculture itself depends on the health of the forest. Crucial watersheds must be guarded in the east. The climate and microclimate reflect the forest cover in west and south. Forestry plantations could also play an important role where the land is now bare — in providing firewood, in preventing erosion, and in putting sterile grassland to good use.

None of this is news to the Eaux et Forets department. They have attempted for many decades to reforest Madagascar, and to conserve the riches of its natural heritage, as explained in the following article by J. Andriamampianina. However, such long-run interests are inevitably sacrificed by a government in financial need, and by millions of small farmers pressed by immediate needs of survival.

Conservation in Madagascar is a tragedy without villains. It can only be turned to success by determined Malagasy policy for the future, supported by concerted foreign aid through this period of crisis. Will the richer countries be wise enough to see that Madagascar needs not just aid for roads, but aid for forests? Are we wise and generous enough to see that aid for forests will only succeed if it eases the pressures on the Malagasy people — which means support for actions which improve living standards, and which in turn must cover a range from improved agricultural markets and fair prices to education and primary health care facilities?

Will the present generation be wise enough to see in our own self-interest that the wonders of Malagasy nature are not just curiosities to embalm in a wildlife book: they are living things? We, both westerners and Malagasy, have a free choice. We may choose to actively guard the riches of our earth, or we may turn aside indifferently, and let the libraries and museums of nature burn.

CHAPTER 14

Nature Reserves and Nature Conservation in Madagascar

JOSEPH ANDRIAMAMPIANINA

THE INTEGRAL NATURE RESERVES

Active protection of nature began long ago in Madagascar. Under the ancient Hova Kingdom, the "Code of 305 articles" of 1881 condemned those who cut down the forests to be chained in irons. We must admit, though, that these rules were hardly respected at any distance from the capital, for there was no forest administration. Forest clearing and burning increased, particularly on the east coast.

At the beginning of this century the progressive disappearance of Madagascar's unique fauna and flora drew the attention of the newly-formed Academie Malgache. From its creation this Institution dedicated a large share of its attention to research on the fauna and flora. Its persistent pressure resulted in the Decree of 31 December 1927, founding the 10 first "Reserves Naturelles Integrales de Madagascar". The International Conference held in London in 1933 later took the Integral Reserves as a model of organization, when it adopted an International Convention for the Protection of African Fauna and Flora.

Later, Professor Humbert proposed two other Integral Reserves, which were created in 1939 and 1952. In 1964, however, one of the 12 reserves was degazetted, in favour of commercial exploitation, which opposes both the purposes and the long-term advantages of these conservation zones.

The Reserves were created with the goal of preserving the diverse types of fauna and vegetation of Madagascar. They should therefore be sanctuaries closed to all human interference, and even to access by the inhabitants of nearby villages. Their sites were chosen to constitute as complete a sample as possible of the various plant communities of the island, and except where there was no other choice, in little populated or mountainous areas which would be shielded from the pressure of a population in constant search of new crop land. Legal access to the Integral Nature Reserves is limited to duly authorized scientists.

The following paragraphs describe some characteristics of each reserve.

Reserve Naturel Integrale No.1: Betampona 2,228 ha.

This reserve is situated Northwest of Toamasina (Tamatave). It is a sample of the natural biotopes of low altitudes in the eastern domain. It is an area of steep topography, including the headwaters of many streams. There are both primary and secondary east coast evergreen rainforests. The fauna includes several lemurs (*Indri indri*, *Hapalemur griseus*, *Phanere furcifer*, etc.), birds and endemic viverrids.

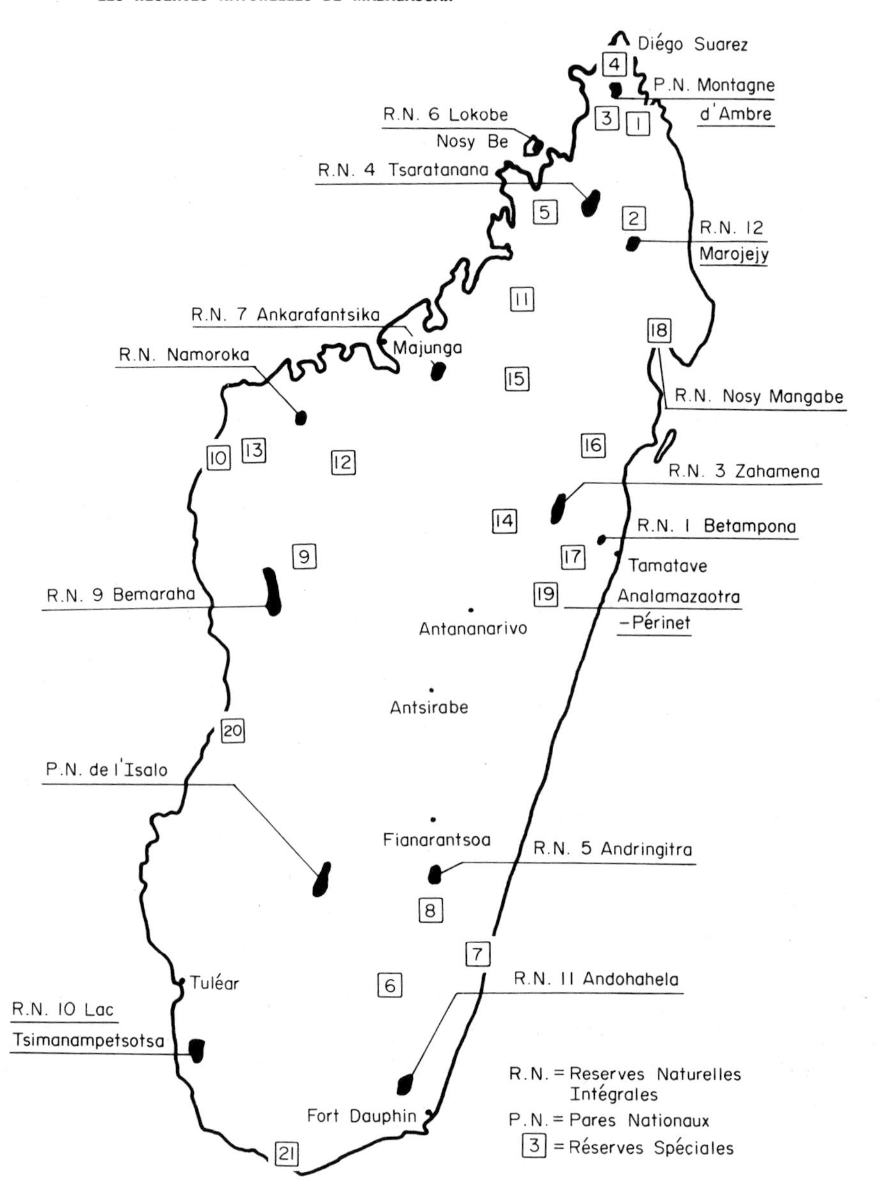

Fig. 14.1 Map of Natural Reserves (Andriamampianina)

R.N.I. 2: Masoala 30,000 ha.

This reserve was degazetted in 1964. It was a rainforest reserve which included nearly virgin coastal forest and successive closed canopy forests from coastal level to elfin woodland at more than 1000 m altitude.

R.N.I. 3: Zahamena 73,160 ha.

Zahamena lies east of Ambatondrazaka. It offers a sample of the vegetation of the eastern escarpment, with primary and secondary tropical evergreen forest, bamboo zones, and montane lichen woodland. Fauna includes many lemurs, (*Indri indri*, *Varecia variegata*, etc.), birds and reptiles.

R.N.I. 4: The Tsaratanana 48,622 ha.

This reserve lies southeast of Ambanja. It protects the highly specialized vegetation of the Tsaratanana massif, which includes Mount Maromokotra, the Island's highest peak (2887 m). There are primary and secondary evergreen forest, at low and high altitudes, lichen woodland and heath. Many orchid species. Fauna includes a few lemurs (*Lemur macaco*, etc.), viverrids, birds and reptiles.

R.N.I. 5: The Andringitra 31,160 ha.

This reserve lies south of Ambalavao, and preserves the flora of the island's second highest point, Pic Boby (2658 m). Many watercourses rise in this area of sharp relief. There are ericoid and herbaceous communities, and xerophytic flora on the rocks (*Aloe*, *Kalanchoe*, *Helichrysum*), with Cunionaceae and Euphorbiaceae among the trees. Fauna with lemurs (*Lemur catta*, *L. fulvus*, *Propithecus diadema*), viverrids (*Cryptoprocta ferox*), birds and reptiles, especially chameleons.

R.N.I. 6: Lokobe 740 ha.

This reserve on the island of Nosy Bé is a sample of low-altitude forest of the Sambirano region. Evergreen primary and secondary forest. This is the only vestige of forest on Nosy Bé, and so plays an important hydrographical role. The fauna includes *Lemur macaco*, *Lepilemur dorsalis* and various reptiles and invertebrates.

R.N.I. 7: The Ankarafantsika 60,520 ha.

This reserve is situated to the east of the main north road, between Maevatanana and Majunga. It has Western deciduous forest on sandy soil, with many Leguminosae and Myrtaceae. The fauna includes the lemurs *Propithecus verreauxi coquereli*, *Lemur fulvus fulvus*, *Lemur mongoz*, *Avahi laniger*, *Lepilemur edwardsi* as well as western birds and reptiles.

R.N.I. 8: The Tsingy of Namoroka 21,742 ha.

This reserve, south of Soalala, has a Western biotope on calcareous rock. The limestone erodes into "tsingy", or vertical lamellae of limestone, with many caves and a deep pool filled by underground springs. There is primary deciduous forest and degraded formations adapted to the karst substrate, with baobabs, xerophiloc plants and succulents. The last remaining ebonies are found there. Fauna includes several endemic birds such as *Coua coquereli*, typical reptiles of the west, and lemurs including *Propithecus verreauxi deckeni*.

R.N.I. 9: The Tsingy of Bemaraha 152,000 ha.

This reserve, east of Antsalova, has a biotope similar to the preceding, but with a greater diversity of species, as well as the cemeteries of early Malagasy cultures.

R.N.I. 10: Lake Tsimanampetsotsa 43,200 ha.

This reserve, south of Tulear, includes the whole of Lake Tsimanampetsotsa, as well as a region covered by the natural vegetation of the Mahafaly Plateau. This is xerophytic bush on calcareous substrate and the typical south western bushy scrub on sandy soils. The flora includes many unique species especially Didiereaceae, a family endemic to the southern region, coralliform Euphorbia trees, Leguminosae, Combretaccae, etc. There are a few lemurs, but this reserve is chiefly important for its avifauna: two species of flamingo, many wading birds, ducks, and other endemic and visiting waterfowl. There are also tortoises of the endemic species *Geochelone radiata*.

R.N.I. 11: Andohahela 76,020 ha.

This reserve, northwest of Fort Dauphin has vegetation of the southern end of the eastern humid forest, of the Mandrare basin, of the southern bush, and transition woodland of the "western slopes". Over only a few kilometres one finds three contrasting vegetation types. There is primary and secondary evergreen rainforest, forests adapted to semi-arid conditions with *Euphorbia* and *Alluaudia*, and a transition zone with *Neodypsis decaryi*, the triangular palm, endemic to the region. Each sector has its own fauna adapted to its own habitat, with diverse birds and lemurs.

R.N.I. 12: Marojejy 60,150 ha.

Northwest of Andapa, this reserve includes the mountain massif of Marojejy, with a continuous altitudinal succession from 100 m to over 2100 m in a very short horizontal distance. There is closed-canopy forest, mountain woodland with lichens, and ericoid bush. There is an abundance of palms, ferns, orchids and balsams. The fauna includes lemurs, birds and insects which are characteristic of the massif.

THE NATIONAL PARKS

The national parks were planned for general education and enjoyment, rather than primarily as conservation areas. Tourists have free access to them, but should still equip themselves with an entry permit from the Direction des Eaux et Forets in Antananarivo.

National Park of the Isalo 81,540 ha.

Lying West of Ihosy, this park includes the grandiose landscape of the Isalo. This is a sandstone massif like ruined castles rising from the plain, with dizzy cliffs, deep gorges, and enclosed canyons. It also contains historic sites of the 16th century (the Grotto of the Portuguese). The flora is highly localized with endemic species like *Pachypodium rosolatum* (Apocynaceae), the palm *Ravenea rivularis*, and *Uapaca bojeri* (Sarcolaenaceae). The fauna includes lemurs like *Lemur catta* and *Propithecus verreauxi verreauxi*, viverrids and endemic reptiles.

National Park of the Montagne d'Ambre 18,200 ha.

This park lies south of Diego Suarez. It includes the Montagne d'Ambre, a mountain covered in evergreen rainforest isolated by drier woodland on the surrounding plains. Crater lakes, waterfalls, and magnificent outlook points with views of the sea add to the park's beauty. The flora is rich in tree ferns, epiphytes and orchids. The lemurs include *Lemur fulvus sanfordi* and *Lemur coronatus*, and there are birds and reptiles. The rare black sifaka *P. diadema perrieri* occurs in the region though not in the park.

THE SPECIAL RESERVES

About twenty special reserves have been created with the goal of completing a sample of biotopes not included in Integral Nature Reserves, or to protect particular animal species. These reserves are, in principle, protected from human exploitation, but certain rights are accorded to surrounding inhabitants to gather plants or fruit for food, or to fish in the watercourses.

Like the national parks and forest stations, these special reserves are accessible to the public under authorization given by the Direction des Eaux et Forets in Antananarivo. We shall describe the reserves of Perinet and Nosy Mangabe, and simply list the others.

Faunal Reserve of Perinet-Analamazotra 810 ha.

This reserve is east of Moramanga. Its essential objective is conservation of *Indri indri*, largest of the Malagasy lemurs. This area has eastern vegetation, in a transition zone between lowland coastal forest and high altitude rainforest, with precious hardwoods, tree ferns and orchids. Besides *Indri indri* there are other lemurs such as *Propithecus d. diadema*, birds, reptiles, viverrids and invertebrates.

Special Reserve of Nosy Mangabe 520 ha.

The little island of Nosy Mangabe, in the Bay of Antongil south of Maroansetra, was elevated to special reserve for the protection of the rarest of Malagasy lemurs, *Daubentonia madagascariensis*, the aye-aye, which is on the way to extinction. Nine specimens of this species were released on the island in 1967.

This island is very steep, with typical lowland east coast forest: *Canarium*, *Ocotea*, *Ravensara*, and various palms and ferns.

List of Special Reserves

The date of creation follows in parentheses:

*Province d'Antsiranana (Diego Suarez):

1. Analamerana, 34,700 ha (1956)
2. Anjanaharibe – Sud, 32,100 ha (1956)
3. Ankara, 18,220 ha (1956)
4. Foret d'Ambre, 4810 ha (1958)
5. Manongarivo, 35,250 ha (1956)

*Province de Fianarantsoa:

6. Kalambatritra, 28,250 ha (1959)
7. Manombo, 5020 ha (1962)
8. Pic d'Ivohibe, 3450 ha (1964)

*Province de Mahajanga (Majunga):

9. Ambohijanahary, 24,750 ha (1958)
10. Bemarivo, 11,570 ha (1956)
11. Bora, 4780 ha (1956)
12. Kasijy, 18,800 ha (1956)
13. Maningozo, 7900 ha (1956)14. Marotandrano, 42,200 ha (1956)
15. Tampoketsa d'Analamaitso, 17,150 ha (1958)

*Province de Toamasina (Tamatave):

16. Ambatovaky, 60,050 ha (1958)
17. Mangerivola, 800 ha (1958)
18. Nosy Mangabe, 520 ha (1965)
19. Analamazaotra-Perinet, 810 ha (1970)

*Province de Toliary (Tulear):

20. Andranomena, 6420 ha (1958)
21. Cap Sainte Marie, 1750 ha (1962)

THE FORESTRY STATIONS

More than thirty forestry stations are concerned with reforestation schemes, arboretums, and water management of natural and artificial reservoirs, which may or may not include areas of natural forest. Some of these, in fact, serve as magnificent national parks where city-dwellers may communicate with nature, in settings of grandeur. Near Tananarive the stations of Angavokely and Manjakatompo are pleasant goals for excursions. We shall mention only the principal Forestry Stations, the most interesting ones for the public to visit.

Forestry Station of Manjakatompo

West of Ambatolampy, this station has a remnant of high altitude forest, a pinetum, limpid rivers where trouts are stocked by the forestry station, hatchery, and lakes filled with waterfowl. Tracks which are passable all year (at least by Land Rover) link various outlook points to the station, almost to the summit of Tsiafajavona, 2643 m high, the uppermost point of the Ankaratra massif. The station also includes the tomb of Andriampenitra, the great ancestor of the Tankaratra. Prayers and sacrifices are held at the tomb, usually at new moon.

Forestry Station of Angavokely

This station lies about 30 km east of the capital, near Carion. It has beautiful landscapes, grottos, archaeological remains, and varied vegetation. There are splendid panoramic views from the three main summits, each a granite monolith.

Forestry Station of Antsampandrano

Lying near Faratsiho, this station is famous for its trout-filled rivers and its pine plantations.

Forestry Station of Mandena

Near Fort Dauphin, this station has vestiges of natural coastal forest with lemurs, birds, and the carnivorous plant *Nepenthes madagascariensis*.

Forestry Station of Ampijoroa

On the road to Majunga, near Ambato-Boeni this station serves Reserve Naturelle Integral No.7. It is often visited by zoologists to observe its lemurs, reptiles and birds.

Station and Botanic Garden of Ivoloina

This station with its arboretum and botanic garden, was founded in 1898, 12 km north of Tamatave. One may see there most of the typical east coast plants, and a breeding centre for lemurs.

Industrial Reforestation of the High Matsiatra

This area is 30,000 ha of reforestation in Mexican Pine, as well as artificial lakes, near Fianarantsoa.
Industrial Reforestation of the High Mangoro.

Near Moramanga, this reforestation scheme includes several tens of thousands of hectares of Indo Chinese Pine, which in a few years will begin to feed a second paper pulp mill in Moramanga.

THE PRIVATE RESERVE OF BERENTY

This rapid overview of the Reserves of Madagascar would be incomplete without special mention of the remarkable private reserve created by the de Heaulme family at Berenty, in the Fort Dauphin region. This family which has long lived in the South of Madagascar, where they have created fine sisal plantations, has chosen to totally protect a section of the gallery forest of the Mandrare Valley. Visitors can admire the rich lemur fauna, particularly *Propithecus verreauxi* and *Lemur catta*, and various zoologists have worked there.

THE PROTECTED ANIMALS

A decree of 16 February, 1961, annulled and replaced a series of earlier laws. It defined and listed the most threatened Malagasy species, whose hunting and capture are forbidden. All lemurs are thus protected (since 1927), as well as the dugong, the blind cave fish, and two terrestrial tortoises: *Geochelone radiata*, the radiated tortoise, and *G. yniphora*, the plowshare tortoise.

The boid serpents, the ''Do'', which are precious auxiliaries in man's warfare with rats and other rodents, are protected because their survival was threatened by the use of snakeskin for leather goods. Wholly protected birds include the crested ibis or ''akohoala'', (*Lophotibis cristata*), the little egret, ''vano, vanofotsy'', (*Egretta garzetta dimorpha*), the large egret, ''vanobe, vanofotsy,'' (*Egretta alba melanorhynchos*), and the flamingoes, ''samaka''.

Mammals which do not originate from the Island but are acclimatized are also protected like the European deer introduced to the forest station of Manjakatompo, and the Indochinese deer introduced to Perinet. These animals, alas, have become very rare and difficult to observe because of poaching*.

Finally, export from the country of any wild animal or parts thereof is strictly controlled.

PROTECTION OF NATURE: A FUNDAMENTAL TASK

In spite of the considerable efforts which have already been made, much remains to do to save our inestimable natural heritage. In 1970 the Government of Madagascar and the International Union for the Conservation of Nature held a conference in Antananarivo on the ''Rational Use and Conservation of Nature''. The success of this meeting was largely due to the efforts of M. Georges Ramanantsoavina, M. Jean-Jacques Petter and Mme Monique Pariente. The Conference made various recommendations, especially:

- strict protection of the final vestiges of virgin lowland forest around Maroansetra;
- increase in the number and extent of the forest reserves;
- intensive reforestation;
- protection of wetlands where many waterbirds live (Lakes Ihotry, Itasy, Kazanga, Kinkony, and the region included between Antsalova, Bekopaka, the Manambolo River and the sea);
- protection of littoral zones, in particular the Great Reef of Tulear and the Bay of Assassins;
- increased protection for sea turtles;

* With New Zealand as an example, we may be grateful to the poachers. — eds.

— protection of *Hypogeomys* the giant jumping rat, *Uratelornis*, the longtailed ground roller, and *Galidictis*, least known of Malagasy viverrids;
— reinstatement of the former Reserve Naturelle Integrale No. 2 of Masoala;
— creation of a National Park in the Sept Lacs region near Tulear;
— creation of a National Park in a representative area of Didiereacea bush.

Several of these recommendations have already been followed up. Statues currently cover Lakes Kinkony (near Mitsinjo), Ihotry (near Morombe), Kazanga (near Soavinandriana), Masamba and Bemamba (near Antsalova).

Still on the legislative side, Madagascar has ratified the African Convention for Conservation of Nature and Natural Resources, and the Convention on International Trade in Endangered Species.

Much remains to do however. One must not surrender to self-deception: in spite of all these laws, protection is only relative. We cannot watch everywhere. The forests are few in numbers, dispersed, and burdened with obligations which seem more urgent or more important for development in the short term.

The work of nature conservation is not a simple matter of prohibitions and laws. The policy of the present Government is to support these laws by popular education. How can we blame a peasant who cuts the forest to replace it by a rice field that will last at most three years, if no one has ever told him that the forests of Madagascar are a scientific treasure for humanity, as well as a source of rain and water for surrounding regions?

Various efforts for public education have already been attempted: radio programs, discussions in villages and youth groups, slogans and posters against forest clearance and bushfires, brochures on fauna and flora. Academic conferences also help transmit information to the public, such as the International Colloquium on Medicinal Plants.

The chief hope for the future in this task of nature conservation lies in the development of national pride, which is happening now. There is a strong current of feeling which emphasises the original and unique aspects of our country. Madagascar's natural heritage is becoming a source of pride, along with Malagasy historical, literary and artistic values. Besides this there is a strong international movement in support of national conservation efforts. The recent creation of a Malagasy Section of the World Wildlife Fund will make a great contribution to these efforts, whose final goal is to hand on to future generations this great heritage — preserved intact — the fauna and flora of Madagascar.

BIBLIOGRAPHY

Andriamampianina, J. 1971 *La protection de la nature a Madagascar*, Antananarivo, 51p., multic.

Andriamampianina, J. 1978. La protection des Lemuriens, Communication au Colloque scientifique international sur les Lemuriens. *Acad. Malg.*

Defenders of Wildlife, 1975 Numero special de cette revue, consacre a la Nature malgache et a sa protection, nombreuses ill. couleurs, vol. 50, n° 2, avr. (2000 N. Street, North-West, Washington DC 20036).

Paulian, R. 1955 *Les Animaux proteges de Madagascar*, Tananarive, I.R.S., 60p.

Revue de Madagascar, 1950 Numero special de cette revue, consacre a la Foret malgache et aux Reserves Naturelles, n° 8, 1er trim. 1950, 47p.

Union Internationale pour la Conservation de la Nature (U.I.C.N.), Compte rendu de la Conference internationale sur la conservation de la nature et de ses ressources a Madagascar, 1972, 273p.

Verin, P. and Griveaud, P. 1968. La protection des richesses naturelles, archeologiques et artistiques a Madagascar. Universite de Madagascar, *Guide d'initiation active au developpement*, 109p.

Fig. 14.2 The Travellers' Palm (W. Ellis, 1865)

Index

KEY ENVIRONMENTS

Other Titles in the Series

AMAZON RAIN FOREST Edited by: G. Prance and T. Lovejoy

ANTARCTICA Edited by: N. Bonner and D. Walton

GALAPAGOS Edited by: R. Perry

MADAGASCAR Edited by: A. Jolly, R. Albignac and P. Oberlé

SAHARA DESERT Edited by: J. L. Cloudsley-Thompson

WESTERN MEDITERRANEAN Edited by: R. Margalef